Undergraduate Topics in Computer Science

Series Editor

Ian Mackie, University of Sussex, Brighton, UK

Advisory Editors

Samson Abramsky ⓘ, Department of Computer Science, University of Oxford, Oxford, UK

Chris Hankin ⓘ, Department of Computing, Imperial College London, London, UK

Mike Hinchey ⓘ, Lero—The Irish Software Research Centre, University of Limerick, Limerick, Ireland

Dexter C. Kozen, Department of Computer Science, Cornell University, Ithaca, NY, USA

Hanne Riis Nielson ⓘ, Department of Applied Mathematics and Computer Science, Technical University of Denmark, Kongens Lyngby, Denmark

Steven S. Skiena, Department of Computer Science, Stony Brook University, Stony Brook, NY, USA

Iain Stewart ⓘ, Department of Computer Science, Durham University, Durham, UK

Joseph Migga Kizza, Department of Engineering and Computer Science, University of Tennessee at Chattanooga, Chattanooga, TN, USA

Roy Crole, School of Computing and Mathematics Sciences, University of Leicester, Leicester, UK

Elizabeth Scott, Department of Computer Science, Royal Holloway University of London, Egham, UK

'Undergraduate Topics in Computer Science' (UTiCS) delivers high-quality instructional content for undergraduates studying in all areas of computing and information science. From core foundational and theoretical material to final-year topics and applications, UTiCS books take a fresh, concise, and modern approach and are ideal for self-study or for a one- or two-semester course. The texts are authored by established experts in their fields, reviewed by an international advisory board, and contain numerous examples and problems, many of which include fully worked solutions.

The UTiCS concept centers on high-quality, ideally and generally quite concise books in softback format. For advanced undergraduate textbooks that are likely to be longer and more expository, Springer continues to offer the highly regarded *Texts in Computer Science* series, to which we refer potential authors.

Gerard O'Regan

Ethical and Legal Aspects of Computing

A Professional Perspective from Software Engineering

Gerard O'Regan
Cork, Ireland

ISSN 1863-7310　　　　　　ISSN 2197-1781　(electronic)
Undergraduate Topics in Computer Science
ISBN 978-3-031-52663-3　　　ISBN 978-3-031-52664-0　(eBook)
https://doi.org/10.1007/978-3-031-52664-0

© The Editor(s) (if applicable) and The Author(s), under exclusive license to Springer Nature Switzerland AG 2024

This work is subject to copyright. All rights are solely and exclusively licensed by the Publisher, whether the whole or part of the material is concerned, specifically the rights of translation, reprinting, reuse of illustrations, recitation, broadcasting, reproduction on microfilms or in any other physical way, and transmission or information storage and retrieval, electronic adaptation, computer software, or by similar or dissimilar methodology now known or hereafter developed.
The use of general descriptive names, registered names, trademarks, service marks, etc. in this publication does not imply, even in the absence of a specific statement, that such names are exempt from the relevant protective laws and regulations and therefore free for general use.
The publisher, the authors, and the editors are safe to assume that the advice and information in this book are believed to be true and accurate at the date of publication. Neither the publisher nor the authors or the editors give a warranty, expressed or implied, with respect to the material contained herein or for any errors or omissions that may have been made. The publisher remains neutral with regard to jurisdictional claims in published maps and institutional affiliations.

This Springer imprint is published by the registered company Springer Nature Switzerland AG
The registered company address is: Gewerbestrasse 11, 6330 Cham, Switzerland

Paper in this product is recyclable.

To
My wonderful godson, Cian Mullane
(To mark the occasion of his 21st birthday)

Preface

Overview

The objective of this book is to give the reader an overview of ethical and legal issues that arise in the computing field, and the book was influenced by the author's experience working as a software engineer at Motorola in Cork, Ireland. His responsibilities at Motorola included software process improvement and software quality management, but he also gained exposure to ethical and legal aspects of computing while coordinating the patent programme, and working with several software suppliers. This motivated him to explore these areas in more detail, especially in their applications to computer science, and so while the author is not a specialist in law or ethics (i.e. the book is written from the point of view of a software engineer who is a non-specialist), he hopes that the reader will find the coverage to be both interesting and insightful.

Organisation and Feature

The introduction of modern computer technology has led to immense changes in society, and the Internet and World Wide Web as well as personal devices such as smartphones and tablets dominate the modern world. The new technologies have impacted on security and privacy with monitoring of online behaviour for commercial purposes. Social media is a useful tool for communication, but it is also an addictive technology and needs to be managed appropriately. Modern technology has led to scams, viruses and computer crime, and individuals need to exercise caution to avoid becoming victims of crime.

The focus of the book is more on the ethical side of computing rather than on the legal side (roughly a 60:40 split), and the early chapters focus on the history of ethics; professional responsibility of computer professionals; ethical software engineering; the ethics of data science; social media and ethics; and AI and ethics. The emphasis then changes to the legal side with chapters on an introduction to law; legal and ethical aspects of outsourcing; intellectual property law; e-commerce law and concludes with a chapter on computer crime.

Chapter 1 introduces ethics and the law, and we discuss ethical and legal problems that arise in the computer field. Chapter 2 presents a history of ethics, and we discuss ethics in ancient civilisations as well as ethics in religious traditions. We discuss utilitarian and deontological ethics and consider the question of what it means to be an ethical person in a complex world.

Chapter 3 discusses the professional responsibilities of computer professionals, and we discuss the code of ethics of the Association for Computing Machinery, the Institute of Electrical and Electronic Engineers and the British Computer Society. We discuss the role of the whistle-blower and its importance.

Chapter 4 discusses ethical software engineering, and we discuss notable failures such as the space shuttle disaster and the defective Therac-25 radiotherapy machine. We discuss the extraordinary Volkswagen emissions scandal, where engineers designed a "defeat device" that would allow Volkswagen cars to pass emission tests in the USA.

Chapter 5 discusses the ethics of data science. There has been a phenomenal growth in the use of digital data with vast amounts of data collected, processed and used, and so the ethics of data science has become important. It aims to investigate what is fair and ethical in data science, and what should or should not be done with data. We discuss ethical problems of privacy that arise in social media, the Internet of Things, AI and facial recognition, and the legal aspects of privacy.

Chapter 6 discusses social media and ethics, and we discuss the Facebook revolution and its impact during the Arab spring as well as social media campaigns. We discuss the Cambridge Analytica affair and its impact in influencing voters in the 2016 election in the USA. We discuss the challenge of fake news, as well as the responsibilities of social media companies to manage this appropriately.

Chapter 7 discusses ethics and AI, and we discuss Weizenbaum's Eliza programme and the challenge that AI poses to human dignity. We discuss the ethics of self-driving cars, and the need to encode an appropriate moral compass in a self-driving car to deal with situations where ethical decisions need to be made. We discuss ethical problems that arise with AI and surveillance as well as ethical problems with expert systems.

Chapter 8 provides a brief introduction to law, where the laws of a country apply to all citizens and residents of the state. Roman law influenced the development of civil law, where the civil law tradition is focused on reasoning on the basis of rules. Sharia law consists of a set of duties that all Muslims are expected to observe, and the law acts as a path to guide Muslims in their relationships with their neighbours, the state and with God. English common law operates on the principle of binding precedent, where the judge in a particular case must follow the decision of judges that have ruled on similar cases in the past. We discuss European law and the European Convention of Human Rights.

Chapter 9 discusses legal and ethical aspects of outsourcing of software development and/or testing. Advanced economies have many laws and regulations to protect the environment and the health and safety of employees in the workplace.

However, the laws and regulations in the country where the subcontractor is based may not be as stringent, and so an ethical corporate citizen must do more than complying fully with the laws of every country it is operating, as these laws may not be fit for purpose. We discuss legal aspects of failure including lawsuits and the law of tort.

Chapter 10 discusses intellectual property including patents, copyrights and trademarks. Intellectual property law deals with the rules that apply in protecting inventions, designs and artistic work, and in enforcing such rights. Patents protect innovative ideas and concepts and give inventors exclusive rights to their invention for a specified period or time. A copyright applies to original writing, music and other original intellectual and artistic expressions. It protects the expression of the idea and not the underlying idea itself. A trademark protects names or symbols that are used to identify goods or services, and their purpose is to avoid confusion and to help customers to distinguish one brand from another.

Chapter 11 discusses e-commerce and e-commerce law, where e-commerce is a way to conduct business online. It generally involves listing products for sale in a catalogue format on a website, and there are also legal issues to consider. E-commerce is different from traditional business in that the buyer and seller do not physically come together in the market place to perform the transaction, and they may even be in different countries with distinct laws. There may be a greater risk of fraud or loss with e-commerce than with traditional commerce.

Chapter 12 discusses computer crime, where a computer crime is a crime that involves a computer and a network. The computer may be the vehicle by which the crime was conducted, or it may be the target of the crime. New laws are introduced to deal with computer crime as technology evolves to allow the legal system to catch up with problems introduced with new technology. The Internet and World Wide Web have transformed computer crime to the global stage, and this has created a whole new set of problems and challenges for the legal system to deal with. We discuss various computer crimes as well as cybersecurity, which plays a key role in their prevention.

The epilogue summarises the journey that we have travelled in this book.

Audience

The main audience of this book are computer science students who are interested in learning about ethical and legal issues in the computing field. It will also be of interest to computer professionals including software engineers, software managers, as well as the general reader.

Acknowledgments

I am deeply indebted to family and friends who supported my efforts in this endeavour, and my thanks, as always, to the team at Springer. This book is dedicated to my godson, Cian Mullane, to mark the occasion of his recent 21st birthday. Cian has shown an early talent for business, and I wish him every success and happiness in future.

In all probability, this is my last Springer book, and so it is very appropriate at this time to pay a special thanks to my editor, *Wayne Wheeler*, who I have worked with for many years (more years than I would care to admit). Wayne was a true professional at all times, and he gave every book proposal a very fair and balanced consideration. He was a pleasure to work with and always gave valuable professional feedback on each potential book project. And so, as a result, I think that we produced some reasonable books over the years, and it was, for me, a very positive working relationship, with nice warm communication over the years. And so, it is important to say:

Go raibh míle maith agat.

Cork, Ireland   Gerard O'Regan

Contents

1	**Introduction to Ethical and Legal Aspects of Computing**		1
	1.1	Introduction	1
	1.2	Ethics	2
	1.3	Business Ethics	4
	1.4	What Are Computer Ethics?	6
		1.4.1 Ethical Problems in Computing	8
		1.4.2 The Digital Divide	8
	1.5	Ethical Software Engineering	11
		1.5.1 The Ethical Software Engineer	12
		1.5.2 The Ethical Software Tester	13
		1.5.3 The Ethical Project Manager	13
	1.6	Legal Aspects of Computing	14
		1.6.1 Intellectual Property	16
		1.6.2 Software Licenses	17
		1.6.3 Legal Aspects of Outsourcing	19
		1.6.4 Tort in Software Engineering	21
		1.6.5 Lawsuits in Computing	22
	1.7	Computer Crime	23
		1.7.1 Hacking	25
	1.8	Review Questions	26
	1.9	Summary	26
	References		27
2	**A Short History of Ethics**		29
	2.1	Introduction	29
	2.2	Ethics in Ancient Civilisations	30
	2.3	Ethics in Ancient Greece	32
	2.4	Ethics in Religious Traditions	35
		2.4.1 Christian Ethics	35
		2.4.2 Islamic Ethics	38
		2.4.3 Buddhist Ethics	41
	2.5	Traditional Chinese Ethics	43
	2.6	Utilitarianism	44

	2.7	Deontological Ethics	45
	2.8	Libertarianism	48
	2.9	Virtue Ethics	48
	2.10	How Can an Individual Be Ethical in a Complex World?	49
	2.11	How Can an Ethical Environment Be Created in an Organisation?	50
		2.11.1 Ethical Decision Making	51
	2.12	Review Questions	52
	2.13	Summary	52
		References	53
3	**Professional Responsibility of Computer Professionals**		55
	3.1	Introduction	55
	3.2	What is a Code of Ethics?	56
	3.3	Role of a Whistle Blower	59
	3.4	IEEE Code of Ethics	61
	3.5	British Computer Society Code of Conduct	61
	3.6	ACM Code of Professional Conduct and Ethics	63
	3.7	Precautionary Principle	66
	3.8	Case Study on Workplace Ethics	67
	3.9	Review Questions	67
	3.10	Summary	67
4	**Ethical Software Engineering**		69
	4.1	Introduction	69
	4.2	Safety and Ethics	73
		4.2.1 Therac-25 Disaster	74
		4.2.2 Space Shuttle Challenger Disaster	76
	4.3	Ethical Project Management	78
	4.4	Ethical Software Design and Development	79
		4.4.1 Volkswagen Emissions Scandal	83
	4.5	Ethical Software Testing	85
	4.6	Review Questions	86
	4.7	Summary	87
		References	87
5	**Ethical and Legal Aspects of Data Science**		89
	5.1	Introduction	89
	5.2	Data Science	90
	5.3	Ethics of Data Science	92
		5.3.1 Problems in Data Science	93
		5.3.2 Problems with Data Science Algorithms	96
	5.4	Data Analytics	97
		5.4.1 Business Analytics and Business Intelligence	98
		5.4.2 Big Data and Data Mining	99

		5.4.3	Data Analytics for Social Media	100
		5.4.4	Mathematics Used in Data Analytics	100
	5.5	Data Privacy		101
		5.5.1	What is Privacy?	104
		5.5.2	Social Media	107
		5.5.3	Internet of Things	108
		5.5.4	AI and Facial Recognition	110
		5.5.5	Privacy and the Law	111
		5.5.6	EU GDPR Privacy Law	112
		5.5.7	EU Digital Services Act	113
	5.6	Security of Data		114
	5.7	Review Questions		115
	5.8	Summary		116
	References			116
6	**Ethical Social Media**			117
	6.1	Introduction		117
	6.2	The Facebook Revolution		119
		6.2.1	The Arab Spring	120
		6.2.2	Social Media Campaigns	124
		6.2.3	Facebook and Privacy	124
		6.2.4	The Cambridge Analytica Affair	125
		6.2.5	Facebook Groups and Moderating Content	127
	6.3	The Tweet		128
	6.4	LinkedIn		130
	6.5	Fake News and Disinformation		131
	6.6	Data Analytics for Social Media		132
	6.7	Ethics and Social Media		133
	6.8	Review Questions		134
	6.9	Summary		135
	References			135
7	**Ethics and AI**			137
	7.1	Introduction		137
	7.2	AI and Human Dignity		140
	7.3	Ethics of Self-driving Cars		141
	7.4	Expert Systems and Ethics		145
	7.5	AI and Unemployment		147
	7.6	AI and Surveillance		149
		7.6.1	Biometric Technology	150
	7.7	AI Algorithms and Discrimination		151
	7.8	AI and Disinformation		152
	7.9	AI and Autonomous Weapon Systems		153
	7.10	Robots and Ethics		155
	7.11	Super-Intelligent Machines		156
	7.12	Rights of Intelligent Machines		157

	7.13	ChatGPT	158
	7.14	Risks and Issues with AI	159
	7.15	Review Questions	160
	7.16	Summary	161
	References		161
8	**Introduction to Law**		**163**
	8.1	Introduction	163
	8.2	English Common Law	166
	8.3	Civil Law	171
	8.4	Sharia Law	173
	8.5	European Law	175
	8.6	European Convention on Human Rights	176
	8.7	Natural Rights and Natural Law	177
	8.8	Human Rights and Human Rights Law	178
	8.9	International Criminal Court	179
	8.10	Freedom of Speech and Responsibility	180
	8.11	Review Questions	180
	8.12	Summary	181
	References		181
9	**Legal and Ethical Aspects of Outsourcing**		**183**
	9.1	Introduction	183
	9.2	Planning and Requirements	186
	9.3	Identifying Suppliers	186
	9.4	Prepare and Issue RFP	187
	9.5	Evaluate Proposals and Select Supplier	187
	9.6	Legal Agreement	188
	9.7	Managing the Supplier	189
	9.8	Acceptance of Software	190
	9.9	Rollout and Customer Support	190
	9.10	Ethical Software Outsourcing	190
	9.11	Legal Breach of Contract	192
	9.12	Review Questions	194
	9.13	Summary	195
	Reference		195
10	**Intellectual Property Law**		**197**
	10.1	Introduction	197
	10.2	Patents	198
		10.2.1 Filing a Patent	202
		10.2.2 Patent Litigation	203
		10.2.3 Case Study—Patent Ruling on First Computer	204

		10.3	Copyright	206
			10.3.1 Copyright for Software	210
			10.3.2 Case Study—Copyright Dispute Apple Versus Microsoft	212
			10.3.3 Software Licenses	214
			10.3.4 Stallman and Free Software	215
		10.4	Trademarks	216
			10.4.1 Legal Aspects of Trademarks	217
			10.4.2 Case Study—Trademark Dispute Apple Corps Versus Apple Computer	219
		10.5	Review Questions	220
		10.6	Summary	221
			References	221
11	Legal and Ethical Aspects of Electronic Commerce			223
		11.1	Introduction	223
			11.1.1 Advantages of E-Commerce	226
			11.1.2 Disadvantages of E-Commerce	227
		11.2	E-Commerce Law	227
			11.2.1 Website and Hosting	229
			11.2.2 Registering a Domain Name	229
			11.2.3 Privacy Policy on Website	230
			11.2.4 Terms and Conditions	231
			11.2.5 Contracts	231
			11.2.6 Sale of Goods or Services	232
			11.2.7 Online Marketing and Advertisement	233
			11.2.8 Jurisdiction	233
			11.2.9 E-Signatures	234
			11.2.10 Payment Systems	234
			11.2.11 Taxation	234
			11.2.12 Consumer Protection	234
		11.3	Ethical Electronic Commerce	235
		11.4	Review Questions	235
		11.5	Summary	235
			References	236
12	Computer Crime			237
		12.1	Introduction	237
		12.2	Types of Computer Crime	239
			12.2.1 Scams	239
			12.2.2 Malware	241
			12.2.3 Credit Card Fraud	242
			12.2.4 Cyberextortion and Ramsonware	243

	12.3	Hacking	244
	12.4	Cybersecurity	246
	12.5	Review Questions	251
	12.6	Summary	251
	References		252

Epilogue 253

Index 255

About the Author

Dr. Gerard O'Regan is an international lecturer in Maths/Computing with research interests in software quality and software process improvement, mathematical approaches to software quality and the history of computing. He is the author of several books in the mathematics and computing fields with Springer.

Abbreviations

ABC	Atanasoff Berry Computer
ACM	Association for Computing Machinery
AECL	Atomic Energy of Canada Ltd.
AI	Artificial Intelligence
AWS	Autonomous Weapon Systems
B2B	Business to Business
B2C	Business to Consumer
B2G	Business to Government
BCS	British Computer Society
BDS	Boycott, Divestment, Sanctions
BI	Business Intelligence
CEI	Computer Ethics Institute
CEO	Chief Executive Officer
ChatGPT	Chat Generative Pre-trained Transformer
CMMI®	Capability Maturity Model Integration
CMM®	Capability Maturity Model
CPU	Central Processing Unit
CSC	Computer Sciences Corporation
CSR	Corporate Social Responsibility
CTO	Chief Technical Officer
CVV	Card Verification Value
DMA	Digital Markets Act
DMCA	Digital Millenium Copyright Act
DoS	Denial of Service
DPC	Data Protection Commissioner
DPIA	Data Privacy Impact Assessment
DPP	Director of Public Prosecutions
DRM	Digital Rights Management
DSA	Digital Service Act
DSC	Digital Services Coordinator
DSL	Digital Subscriber Line
DVD	Digital Versatile Disc
ECHR	European Convention on Human Rights
ECSC	European Coal and Steel Community

EDVAC	Electronic Discrete Variable Automatic Computer
EEC	European Economic Community
ENIAC	Electronic Numeric Numerical Integrator and Computer
EU	European Union
EULA	End User License Agreement
FDA	Food and Drug Administration
FIPs	Fair Information Processing principles
FOSS	Free Open-Sourced Software
FSF	Free Software Foundation
GDP	Gross Domestic Product
GDPR	General Data Protection Regulation
GNU	GNU's Not Unix!
GPL	General Public License
GPS	Global Positioning System
GUI	Graphical User Interface
HR	Human Resources
HSE	Health Service Executive
HTML	Hyper Text Mark-up Language
HTTP	Hyper Text Transfer Protocol
IBM	International Business Machines
ICC	International Criminal Court
ICCPR	International Covenant on Civil and Political Rights
IEC	International Electrotechnical Commission
IEEE	Institute of Electrical and Electronic Engineers
IHRL	International Human Rights Law
IoT	Internet of Things
IP	Internet Protocol
IPO	Initial Public Offering
ISEB	Information System Examination Board
ISO	International Standards Organization
ISP	Intermediary Service Provider
ISP	Internet Service Provider
ISTQB	International Software Testing Qualification Board
JAD	Joint Application for Development
KLOC	Thousand Lines of Code
KPI	Key Performance Indicator
LAWS	Lethal Autonomous Weapon Systems
LFR	Live Facial Recognition
LGBT	Lesbian, Gay, Bisexual, Transgender
NASA	National Aeronautics and Space Administration
NATO	North Atlantic Treaty Organization
OSP	Online Service Provider
OSS	Open-Source Software
PARC	Palo Alto Research Centre
PLATO	Programmed Logic for Automatic Teaching Operations

PMI	Project Management Institute
RFP	Request for Proposal
RUP	Rational Unified Process
SaaS	Software as a Service
SDI	Strategic Defence Initiative
SEC	Securities and Exchange Commission
SEI	Software Engineering Institute
SIG AI	Special Interest Group AI
SIG SOFT	Special Interest Group Software Engineering
SLA	Service-Level Agreement
SNS	Social Networking Site
SOW	Statement of Work
SRB	Solid Rocket Booster
SSL	Secure Socket Layer
TCP	Transmission Control Protocol
TDI	Turbo-charged Direct Injection
TEU	Treaty of the European Union
TFEU	Treaty Functioning of the European Union
TPV	Third-Party Verification
UAT	User Acceptance Testing
UDHR	Universal Declaration Human Rights
UK	United Kingdom
UML	Unified Modelling Language
UN	United Nations
UNCTAD	United Nations Conference on Trade and Development
VCD	Valued Centred Design
VLOP	Very Large Online Platform
VLOSE	Very Large Online Search Engine
VPN	Virtual Private Network
WHO	World Health Organisation
ZB	Zetta Byte

List of Figures

Fig. 1.1	Corrupt legislation 1896. Public domain	4
Fig. 1.2	Legal contract. Creative Commons	20
Fig. 1.3	Dandy pickpockets (1818)	23
Fig. 2.1	Code of Hammurabi Stele. Louvre Museum. Creative Commons	31
Fig. 2.2	Plato and Aristotle	34
Fig. 2.3	The crucifixion of Christ	37
Fig. 2.4	The Ka'baa in Mecca, Saudi Arabia	39
Fig. 2.5	Buddha teaching Four Noble Truths	42
Fig. 2.6	Confucius, Tang Dynasty	44
Fig. 2.7	Immanuel Kant	46
Fig. 3.1	Whistle blower	58
Fig. 4.1	A radiotherapy machine	74
Fig. 4.2	Space challenger disaster	77
Fig. 4.3	Bridge over the River Kwaii in Kanchanaburi	80
Fig. 4.4	Balancing an ethical life against a feather in Egyptian Religion	82
Fig. 4.5	Volkswagen beetle type 82E	84
Fig. 5.1	Bentham's Panopticon prison	102
Fig. 5.2	Cardinals eavesdropping in the Vatican	105
Fig. 5.3	Young peoples on smart phones and social media. Public Domain	107
Fig. 5.4	Fitbit surge. Smart-watch activity tracker. Creative Commons	109
Fig. 5.5	EU GDPR 2016/679	112
Fig. 6.1	Mark Zuckerberg	119
Fig. 6.2	Arab Spring in Middle East	121
Fig. 6.3	Cambridge Analytica Scandel	127
Fig. 6.4	Jack Dorsey at the 2012 time 100 Gala	128
Fig. 7.1	Eliza program	142
Fig. 7.2	Waymo self-driving Car in California, 2017	142
Fig. 7.3	Trolley problem	143

Fig. 7.4	Expert (Albert Einstein)	146
Fig. 7.5	Unemployed in Chicago in 1930s	148
Fig. 7.6	Big Brother is watching you	149
Fig. 7.7	TOSY Ping Pong playing robot 2009	158
Fig. 8.1	King John signs Magna Carta	168
Fig. 8.2	Separation of powers in US Constitution	170
Fig. 8.3	Sharia family law court case	173
Fig. 8.4	Al Azhar University. Cairo	175
Fig. 8.5	Flag of European Union	176
Fig. 8.6	Eleanor Roosevelt holding the Universal Declaration Human Rights	178
Fig. 9.1	Legal contract	188
Fig. 10.1	Patent for an invention	199
Fig. 10.2	Replica of ABC computer: creative commons	205
Fig. 10.3	St. Colomba's Cathach	207
Fig. 10.4	Richard Stallman at Pittsburgh University	215
Fig. 11.1	E-Commerce transaction	224
Fig. 12.1	Trojan horse at Troy	242
Fig. 12.2	Hacker at work on a blacklit keyboard. Creative Commons	245
Fig. 12.3	Cybersecurity strategy. Creative Commons	248
Fig. 12.4	Sun Tzu Wu	249

List of Tables

Table 1.1	Ten commandments on computer ethics	7
Table 1.2	A selection of ethical problems in computing	9
Table 1.3	Some areas of law in computing	15
Table 1.4	Types of Lawsuits	19
Table 2.1	Combinations of legal and ethical	52
Table 3.1	Professional responsibilities of software engineers	56
Table 3.2	Types of professional codes	57
Table 3.3	Steps in whistle blowing	60
Table 3.4	IEEE code of ethics	62
Table 3.5	BCS code of conduct	63
Table 3.6	ACM code of conduct	64
Table 5.1	Sources of information	91
Table 5.2	Questions on data collection	94
Table 5.3	Types of data analytics	97
Table 5.4	Mathematics in data analytics	101
Table 5.5	Principles of data collection	106
Table 5.6	Threats in social media	108
Table 7.1	Challenges with driverless vehicles	143
Table 7.2	Expert systems	145
Table 7.3	Asimov's laws of robotics	155
Table 7.4	Advantages of robots	156
Table 7.5	Disadvantages of robots	156
Table 7.6	Risks and issues with AI	160
Table 8.1	Systems of Law	166
Table 9.1	Supplier selection and management	185
Table 9.2	Possible breaches of contract	194
Table 10.1	Process for obtaining a patent	201
Table 10.2	Exceptions to copyright	209
Table 11.1	Characteristics of E-commerce websites	225
Table 12.1	Examples of computer crime	240

Introduction to Ethical and Legal Aspects of Computing

Abstract

This chapter introduces ethics and the law and we discuss ethical and legal problems that arise in the computer field.

Keywords

Ethics • Ethical decision making • Law of tort • Professional responsibility • Privacy • Test outsourcing • Software licenses • Computer crime • Hacking

1.1 Introduction

The objective of this book is to give the reader an overview of ethical and legal issues that arise in the computing field, and the book was influenced by the author's experience working as a software engineer at Motorola in Cork. His responsibilities at Motorola included software quality management and software process improvement, but he also gained exposure to ethical and legal aspects of computing while coordinating the patent programme at the Motorola plant, and working with several software suppliers of the company. This motivated him to explore these areas in more detail, especially in their applications to computer science, and so while the author is not a specialist in ethics or the legal field (i.e., the book is written from the point of view of a non-specialist) he hopes that that the reader will find the coverage to be both interesting and insightful.

The focus of the book is more on the ethical side rather than on the legal side (roughly a 60:40 split), and the early chapters focus on the history of ethics; professional responsibility of computer professionals; ethical software engineering; the ethics of data science; social media and ethics; and AI and ethics. The emphasis then changes to the legal side with chapters on an introduction to law; legal and ethical aspects of outsourcing; intellectual property law; e-commerce law and concludes with a chapter on computer crime.

Ethics explore what actions are right or wrong within a specific context or within a certain society, and seek to find satisfactory answers to moral questions. It is a search for moral principles to guide the behaviour of individuals or groups, and ethical issues occur when a conflict arises between an individual's moral compass, and the values or moral principles held by the society or organisation that the individual belongs to.

Legal aspects of computing are concerned with the application of the legal system to the computing field. This includes the protection of intellectual property such as patents, copyright, trademarks and trade secrets, and the resolution of disputes that arise between two or more parties. Software outsourcing is where an organisation outsources all or part of the software development to another party, and this is accompanied by a legal contract between the parties. Legal aspects of electronic commerce is concerned with the rules and regulations for purchasing online goods and services.

The legal system consists of a set of laws and rules that guides human behaviour by permitting some actions and forbidding others. Laws are generally made by the legislature of the state, and a particular act is either permitted or not. There are consequences (enforced by the state) for those who do not follow the rules of the state. A good law is generally moral but a law has no necessary basis in morality, and immoral laws could be part of the legal system. In democratic societies laws are generally moral, as citizens have the right to demonstrate against unjust and immoral laws, and governments that fail to respond accordingly would rapidly lose their legitimacy and be removed from power at the next election. However, in totalitarian or autocratic states there may be risks of arrest or imprisonment for protesters, and so the citizens may have no choice but to accept and obey the laws of the state irrespective of whether they are just or not.

There are two broad classes of the legal system namely criminal law and civil law. Criminal law refers to the laws governing crimes against the state, whereas civil law is used to resolve disputes that arise between two or more parties. There is a different burden of proof required in criminal cases (proof beyond reasonable doubt), as the consequence of a successful criminal case taken by the state may be the deprivation of the liberty of an individual.

1.2 Ethics

Ethics is a practical branch of philosophy that deals with moral questions such as what is right or wrong, and how should a person respond when presented with a situation that requires ethical decision making to proceed? Ethics explore what actions are right or wrong within a specific context or within a certain society, and seek to find satisfactory answers to moral questions. It is a search for moral principles to guide the behaviour of individuals or groups, and ethical issues occur when a conflict arises between an individual's moral compass, and the values or moral principles held by the society or organisation that the individual belongs to. The origin of the word "ethics" is from the Greek word ἠθικός, which means habit or custom.

There are various schools of ethics such as the *relativist* position (as defined by Protagoras), which argues that each person decides on what is right or wrong for him; *cultural relativism* argues that the particular society determines what is right or wrong based upon its cultural values; *deontological ethics* (as defined by Kant) argues that there are moral laws to guide people in deciding what is right or wrong and individuals have a duty to follow the moral law; and *utilitarianism* which argues that an action is right if its overall affect is to produce more happiness than unhappiness in society.

Professional ethics are a code of conduct that governs how members of a profession deal with each other and with third parties. A professional code of ethics expresses ideals of human behaviour, and it defines the fundamental principles of the organisation, and is an indication of its professionalism. Several organisations such as the Association Computing Machinery (ACM), the Institute of Electrical and Electronic Engineers (IEEE) and the British Computer Society (BCS) have developed a code of conduct for their members, and violations of the code by members are taken seriously and are subject to investigations and disciplinary procedures (see Chap. 3). Groucho Marx[1] had a humorous quote on principles (or lack of principles).

> Those are my principles, and if you don't like them ... well, I have others.

Business ethics define the core values of the business, and are used to guide employee behaviour. Should an employee accept gifts from a supplier to a company as this could lead to a conflict of interest? A company may face ethical questions on the use of technology. For example, should the use of a new technology be restricted because people can use it for illegal or harmful actions as well as beneficial ones?

An example is mobile phone technology, which has transformed communication between people, and thus is highly beneficial to society. What about mobile phones with cameras? On the one hand, they provide useful functionality in combining a phone and a camera. On the other hand, they may be employed to take indiscreet and inappropriate photos of others without their permission, and then placed on inappropriate sites causing distress and harm to others. In other words, how can the benefits of technology be balanced with unintended consequences, and how can individuals be protected from inappropriate use of technology?

Morality does not directly tell us what we should or shouldn't do, and it is in a sense a set of standards to evaluate good or bad behaviour. An individual needs

---

[1] The Marx brothers (Groucho, Chico, Harpo, and Zeppo) were a famous American family comedy act in the 1930s, and they made hilarious movies such as "Duck Soup" and "A Night at the Opera". Groucho was renounced for his quick with and his extremely funny one-liners, and his quote on principles essentially indicates that his character does not live by any principles.

to be conscious of ethical situations that arise in human existence, and to use their moral compass and values to decide on the appropriate response and to do the right thing. Moral standards are important for the proper functioning of society.

The influential Canadian philosopher and theologian, Fr. Bernard Lonergan, outlined four key steps on being human in his influential book "Insight: A study on human understanding" [1]. These steps are based on four levels of conscious functioning namely experience, understanding, judgment and decision. Lonergan summarises the demands of the human spirit by "*Be attentive, be intelligent, be reasonable, and be responsible*".

1.3 Business Ethics

Business ethics (also called corporate ethics) is concerned with ethical principles and moral problems that arise in a business environment (Fig. 1.1). They refer to the core principles and values of the organization, and apply throughout the organization. They guide individual employees in carrying out their daily roles, and they include the rights and duties of a company, its employees, customers and suppliers.

Many corporation and professional organizations have a written "*code of ethics*" that defines the professional standards expected of all employees in the company. Unfortunately, sometimes the code of ethics of an organisation is just window

Fig. 1.1 Corrupt legislation 1896. Public domain

dressing, where it gives the impression that these are the core values of the organisation. However, the reality on the ground may be quite different, with values neither properly implemented on the ground, or not being rigorously followed by employees in their day-to-day work practices. That is, the code of ethics needs to be ingrained in organisation culture.

All employees are expected to adhere to the core values of the code of ethics in their daily performance of their duties and whenever they represent the company. The human resource function in a company plays an important role in promoting ethics, and in putting internal HR policies in place relating to the ethical conduct of the employees, as well as addressing discrimination, sexual harassment and ensuring that employees are treated appropriately (including cultural sensitivities in a multi-cultural business environment). HR has a responsibility to provide training and awareness to staff on its core values, and it needs to investigate any alleged violations of the code of ethics of the company, and discipline employees (including possible termination of employment) for any serious violations of the code of ethics.

Companies are expected to behave ethically and not to exploit their workers. There was an infamous case of employee exploitation at the Foxconn plant (an Apple supplier of the *i*Phone) in Shenzhen in China in 2006, where conditions at the plant were so dreadful (long hours, low pay, unreasonable workload, and crammed accommodation) that several employees committed suicide. The scandal raised questions on the extent to which a large corporation such as Apple should protect the safety and health of the factory workers of its suppliers. Further, given the immense profits that Apple makes from the *i*Phone, is it ethical for Apple to allow such workers to be exploited?

Today, the area of *corporate social responsibility* (CSR) has become applicable to the corporate world, and it requires the corporation to be an ethical and responsible citizen in the communities in which it operates (even at a cost to its profits). It is therefore reasonable to expect a responsible corporation to pay its fair share of tax, and to refrain from using tax loopholes to avoid paying billions in taxes on international sales. Today, environment ethics has become topical, and it is concerned with the responsibility of business in protecting the environment in which it operates. It is reasonable to expect a responsible corporation to make the protection of the environment and sustainability part of its business practices, even if this has an impact on its profitability.

Unethical business practices refer to those business actions that don't meet the standard of acceptable business operations, and they give the company a bad name. It may be that the entire business culture is corrupt or it may be result of the unethical actions of an employee. It is important that such practices be exposed, and this may place an employee in an ethical dilemma (i.e., the loyalty of the employee to the employer versus doing the right thing such as becoming a *whistleblower* and exposing the unethical or unsafe business practices). There are dangers that a whistleblower could suffer career suicide as a result of exposing unethical practices, and organisations need to create an effective structure or mechanism, where employees can raise these serious issues to enable them to be resolved without fear of negative consequences to their career (see Chap. 3).

Some accepted business practices in the workplace might cause ethical concerns. For example, in many companies it is normal for the employer to monitor email and Internet use to ensure that employees do not abuse them, and so employees may have grounds for privacy concerns. On the one hand, the employer is paying the employee's salary, and has a reasonable expectation that the employee will be professional and will not abuse email and the Internet. On the other hand, the employee has reasonable rights of privacy provided that computer resources are not abused.

The nature of privacy is relevant in the business models of several technology companies. For example, Google specializes in Internet based services and products, and its many products include *Google Search* (the world's largest search engine); *Gmail* for email; and *Google Maps* (a web mapping application that offers satellite images and street views). Google's products gather a lot of personal data about its users that is used to create revealing profiles of the users, which can then be exploited for commercial purposes.

A Google search leaves traces on both the computer and in records kept by Google, which has raised privacy concerns as such information may be obtained by a forensic examination of the computer, or in records obtained from Google or the Internet Service Providers (ISP). Gmail automatically scans the contents of emails to add context sensitive advertisements to them and to filter spam, which raises privacy concerns, as it means that all emails sent or received are scanned and read by some computer.

Google has argued that the automated scanning of emails is done to enhance the user experience, as it provides customized search results, tailored advertisements, and the prevention of spam and viruses. Google maps provides location information which may be used for targeted advertisements, and smartphones with Google maps have a location tracking feature that may be used for the electronic surveillance of users by recording the places that they visit as well as the times and duration of the visit.

The surveillance practices of many technology companies have led to the new field of *surveillance capitalism*, where the personal data of individuals is gathered and exploited for commercial purposes. The ethics of data science is discussed in Chap. 5.

1.4 What Are Computer Ethics?

Computer ethics is a set of principles that guide the behaviour of individuals when using computer resources. Several ethical issues that may arise include privacy concerns, intellectual property rights, as well as the impacts of computer technology on wider society. The more awareness and knowledge that computer professionals and the general public has on the ethical implications of computer

1.4 What Are Computer Ethics?

Table 1.1 Ten commandments on computer ethics

No.	Description
1	Thou shalt not use a computer to harm other people
2	Thou shalt not interfere with other people's computer work
3	Thou shalt not snoop around in other people's computer files
4	Thou shalt not use a computer to steal
5	Thou shalt not use a computer to bear false witness
6	Thou shalt not copy or use proprietary software for which you have not paid
7	Thou shalt not use other people's computer resources without authorization or proper compensation
8	Thou shalt not appropriate other people's intellectual output
9	Thou shalt think about the social consequences of the program you are writing or the system you are designing
10	Thou shalt always use a computer in ways that ensure consideration and respect for your fellow humans

technology the better they will be able to navigate through these issues in a thoughtful and considered manner.

The Computer Ethics Institute (CEI) is an American organisation that examines ethical issues that arise in the information technology field. It published the *ten commandments on computer ethics* in the early 1990s [2], which attempted to outline principles and standards of behaviour to guide people in the ethical use of computers (Table 1.1).

The first commandment says that it is unethical to use a computer to harm another user (e.g., destroy their files or steal their personal data), or to write a program that on execution does so. That is, activities such as spamming, malware, spyware, phishing, ransomware, and cyberbullying are unethical.

The second commandment is related and may be interpreted that malicious software and viruses that disrupt the functioning of computer systems are unethical. The third commandment says that it is unethical (with some exceptions such as dealing with cybercrime and international terrorism) to read another person's emails, files and personal data, as this is an invasion of their privacy.

The fourth commandment states that cybercrime and the theft or leaking of confidential electronic personal information is unethical (computer technology has made it easier to commit fraud from the theft of personal information). The fifth commandment states that it is unethical to spread false or incorrect information (e.g., the spreading of fake news or misinformation via email or a social media platform).

The sixth commandment states that it is unethical to obtain illegal copies of copyrighted software, as software is considered an artistic or literary work that is subject to copyright law or license. All copies should be obtained legally. The seventh commandment states that it is unethical to break in to a computer system with another user's id and password (without their permission), or to gain unauthorised access to the data on another computer by hacking into the computer system.

The eight commandment states that it is unethical to claim ownership of an intellectual creation written by another, and so it would be unethical to claim ownership of a program that was written by another as in copyright violation, or to use an invention that is protected by a patent without proper authorisation. The ninth commandment states that it is important for companies and individuals to think about the social impacts of the software that is being created, and to create software only if it is beneficial to society (i.e., it is unethical to create malicious software or addictive software that may harm others). That is, individual and companies need to consider the common good as well as profitability.

The tenth commandment states that communication over computers and the Internet should be courteous, and should show courtesy and respect for others (e.g., there should be no use of abusive language or spreading of false or misleading information).

1.4.1 Ethical Problems in Computing

The ten commandments of computer ethics describe various principles to guide ethical behaviour in the information technology field. The computing field has introduced a unique set of ethical problems such as the unauthorized use of computer resources, the problem of privacy with the gathering of personal data, the development of addictive technology, the problem of hacking and theft of personal data, the problem of computer viruses, the professional responsibility of computer professionals in developing software to the highest standards, the monitoring of employee use of computer resources in the workplace, the protection of personal data and privacy, and computer crime. Some ethical problems that arise in the computing field are summarized in Table 1.2.

1.4.2 The Digital Divide

The Digital Divide refers to the unequal access to digital technology around the world, including access to hardware devices such as smartphones, laptops, and tablets, as well as access to the resources on the Internet and World Wide Web. This means that there is a gap and deep inequalities between people who have access to modern communications technologies and those that don't, as well as a gap between highly developed countries and less developed countries. People without access to modern digital technology are disadvantaged, as they are unable to search for and apply for jobs online, purchase goods or services on line, as well as participating in on-line activities such as social media, obtaining news or researching information via search engines, participating in online conference calls, and so on.

1.4 What Are Computer Ethics?

Table 1.2 A selection of ethical problems in computing

Type	Description
Privacy	The use of computer technology raises concerns on data protection and privacy, as sensitive data may be compromised
Computer crime	This may involve the theft of funds using a computer, or the theft of confidential information through unauthorised access of computer resources
Viruses	A virus is malicious code that an individual places on a network, and it is designed to spread and infect other machines. The virus may have destructive behaviour such as destroying data
Hacking	This is where a hacker who uses his (or her) computer skills to gain unauthorised access to computer files or networks (to cause damage or steal confidential information)
On line monitoring of staff/users	This is where employers monitor the online use of the Internet by staff to ensure that it is not abused, or where companies monitor the online behaviour of users to build up user profiles that may be sold on to advertisers
Fake news	This is where false or misleading information is spread online with the intention of influencing user behaviour or outcomes
Cyberbullying	This is where an individual is bullied online, and it may lead to deep emotional distress to the individual
Facial recognition/surveillance	AI technologies allow faces to be recognised at mass demonstrations, and GPS technology allows for mass surveillance where the location of individuals may be tracked
Deepfakes	AI technologies allow videos to be created that appear to show an individual performing / saying something inappropriate. They have to date been used for comedy/satire but may potentially be used for more sinister purposes in the future

(continued)

Table 1.2 (continued)

Type	Description
Addictive technology	Technology addictions may be dangerous leading to psychological problems and adversely affecting an individual's mental well being as well as their education or career, This includes gaming addiction, social media addiction, on line gambling addiction, pornography addiction, and so on
Professional responsibility of software engineers	The development of a software product is a professional activity, and software engineers have a professional responsibility to ensure that the software product is designed and implemented correctly and adheres to the highest possible standards. Software engineers must be accountable for their decisions and actions, and must ensure that the software is safe to use and does not endanger the public

Inequality of access to the state of the art in technology is not a new phenomenon, as those with visual impairment were unable to access the printed media until the publication of the Braille system (as developed by Louis Braille in France in the 1820s). The Braille system allows the visually impaired to read the news using their sense of touch, with collections of raised dots in a 3 × 2 matrix format representing characters. Access to televisions for the hearing impaired was initially non-existent, but it has been improved with the support for sign language and sub-titles on the screen, and lip reading may be possible in their native language. Monastic communities used early forms of sign languages in Europe from the tenth century, and sign languages have been in use since the sixteenth century in England. The visually and hearing impaired communities have suffered inequalities in their access to education to education throughout history.

The Covid-19 pandemic demonstrated the importance of digital technology, and highlighted the digital divide around the world. The pandemic led to the rapid spread of infection, and so governments around the world were required to impose lockdowns of their population in line with World Health Organisation (WHO) guidelines. This led to major disruptions of everyday life, and meant that non-essential workers could no longer attend the workplace to carry out their employment, and that children and students could not attend school or university to be educated.

It meant that there was a radical need to find alternative approaches to conduct every day life, and the use of the Internet played an essential role in adjusting to this brave new world. It led to the introduction of virtual classrooms for children and students, on-line shopping to purchase groceries and goods, and remote working for non-essential employees. All of these required high-speed broadband, and the digital divide between children from affluent backgrounds and those from lower socio-economic groups was very evident. Often, those from poorer backgrounds had unreliable Internet access in the home, or lacked a computer to complete their assignments.

There are several perspectives by which the digital divide between those that have access to the appropriate digital environment and those that don't may be measured including:

- Digital Infrastructure and participation
- Digital Skills
- Digital Accessibility in Location.

Digital infrastructure refers to the technology that individuals and businesses use to connect to the Internet including computers, laptops and smartphones, as well as the way in which the Internet is accessed such as by fixed or mobile broadband, digital subscriber lines (DSL), and satellites. Digital skills refer to the skills required to make use of the information once the individual has the required technology to access the Internet, and includes skills such as the ability to use electronic mail and search engines, as well as knowledge of how to use Microsoft Office packages such as Word, Excel and Powerpoint. Digital accessibility refers to where the Internet is accessed such as at the library, in schools, in offices and in the home.

The inequality in accessibility of the Internet around the world is evident, as approximately 35% of the world's population does not have access to the Internet in the early 2020s. Further the speed of connection to the Internet is problematic, as in some parts of the world the upload and download speeds to the Internet are not fit for purpose.

1.5 Ethical Software Engineering

Software engineering involves multi-person construction of multi-version programs. It requires the engineer to state precisely the requirements that the software product is to satisfy, and to produce designs that will meet these requirements. It involves starting with a precise description of the problem to be solved; producing a design to solve the problem and validating the correctness of the design; finally, the design is implemented in some programming language, and testing performed to verify its correctness.

The eminent computer scientist, David Parnas, has argued that computer scientists need the right education to apply scientific and mathematical principles in their work. Software engineers need the right engineering education on creating specifications and designs, turning designs into programs, software inspections and testing. The appropriate engineering knowledge should enable the software engineer to produce well-structured programs using module decomposition and information hiding.

Further, Parnas argues that "*software engineers have individual responsibilities as professionals*".[2] They are responsible for designing and implementing high-quality and reliable software that is safe to use. They are also accountable for their own decisions and actions, and have a responsibility to object to decisions that violate professional standards.

Professional engineers have a duty to their clients to ensure that they are solving the real problem of the client. They need to precisely state the problem before working on its solution. Engineers need to be honest about current capabilities when asked to work on problems that have no appropriate technical solution, rather than accepting a contract for something that cannot be done.[3] Ethical software engineering is described in more detail in Chap. 4.

1.5.1 The Ethical Software Engineer

Software engineers have a professional responsibility and are required to behave ethically with their clients, and they may be members of professional bodies such as IEEE, ACM or BCS. The membership of the professional engineering body requires the member to adhere to the code of ethics of the profession. The code of ethics[4] will detail the responsibilities and expected ethical behaviour of the employee including:

- Honesty and fairness in dealings with clients.
- Responsibility for actions.
- Continuous learning to ensure appropriate knowledge to serve the client effectively.

The *licensing of a professional engineer* provides confidence that the engineer has the right education, experience to build safe and reliable products. Otherwise, the profession gets a bad name because of poor work carried out by unqualified people. Professional engineers are required to follow rules of good practice and to object when rules are violated. The licensing of an engineer requires that the engineer completes an accepted engineering course, and understands the professional responsibility of an engineer. The professional body is responsible for enforcing

---

[2] The concept of accountability for actions dates back thousands of years. The ancient Babylonians employed a code of laws c. 1750 B.C. known as 'The Hammarabi Code'. This included a law that if a house collapsed and killed the owner then the builder of the house would be executed.

[3] Parnas applied this professional responsibility faithfully when he argued against the Strategic Defence Initiative (SDI), as he believed that the public (i.e., taxpayers) were being misled and that the goals of the project were not achievable.

[4] These are core values of most mature software companies and many companies today have a code of ethics that employees are required to adhere to. Sometimes the code of ethics is just window dressing.

standards and certification. The term *'engineer'* is a title that is awarded on merit, but *it also places responsibilities on its holder.*

1.5.2 The Ethical Software Tester

Software testers are software engineering professionals and need to behave ethically at all times during the development and testing of the software. The ISTQB Code of Ethics for test professionals is based on the IEEE and ACM code of ethics, and it states that software testers should act in the public interest and in the best interest of their client and employer. They should ensure that their deliverables meet the highest standards, and that they are independent in their professional judgments. They are required to be ethical and to be supportive of their colleagues, and to work closely with software developers. Software testers need to keep their knowledge up to date and so they have a professional responsibility to participate in continuous and lifelong learning.

It is impossible to test everything due to time constraints, and so the testers need to focus their testing on the areas of greatest risk. It is essential that the testers have the appropriate expertise, that the right test environment is set up, that they have prepared appropriate test plans and test specification to test the software, and that they have all the required hardware and tools in place to conduct high quality testing.

Ethical issues could arise during the testing of the software when the project is behind schedule, and where there is pressure applied to the testing team to stay with the original committed project delivery schedule. This could lead to the time available for testing to be compressed (especially in traditional software engineering), resulting in the quality of the released software being compromised. Therefore, the test manager needs to be firm and proactive in resisting any pressure that poses risks to the quality and safety of the software.

1.5.3 The Ethical Project Manager

Project managers need to behave professionally and ethically at all times during the project, and the core values of an ethical project manager includes:

- Professional responsibility
- Respect
- Fairness
- Honesty.

Project management is concerned with the effective management of software projects to ensure the successful delivery of a high-quality product, on time and on budget, to the customer. The project manager is accountable for the success of the project, and larger projects have more opportunities for ethics being compromised than smaller projects. Project managers endeavour to balance budget, schedule, effort and quality, which may potentially lead to ethical dilemmas when the project manager is tempted to cut corners to enable the project to be delivered on time and on budget. This could potentially result in quality being compromised, health and safety being compromised, privacy being compromised, and so on.

The selection of a subcontractor could pose a conflict of interest to the project manager, where the project manager knows one of the candidate subcontractors from a previous working relationship or family relationships. It is therefore important that in such a situation that the project manager excludes herself from the supplier selection to ensure that there is no conflict of interest.

Project management involves ethical decision-making, and good project governance is a good enabler of ethical project management. It enables the key project stakeholders to be kept informed of the key project status and the key decisions being made regularly during the project. Any unethical or illegal conduct should be reported to management, and project management professionals should be aware of the regulations and laws that govern their work.

1.6 Legal Aspects of Computing

Modern society is governed by various rules of behaviour such as rules of etiquette, rules from religion, rules of membership of an organisation, and laws (legal rules) of behaviour. Laws are made by the legislature of the state, and may be enforced by the various organs of the state with fines or imprisonment.

The origin of civil law is from the Roman world, and this is a codified system that specifies the rules and regulations for the purpose of providing civil order in a society. They specify what may be brought to court as well as the applicable procedures and punishment. These laws are generally defined and introduced by legislation in parliaments, and judges interpret the law and the intentions of parliament. They may interpret the law literally or modify the interpretation (e.g., extending the definition in a statue or considering what problem the legislators were attempting to solve) to prevent absurd results. The role of the judge is to establish the facts and to apply the applicable code.

English common law developed in England from the twelfth century, when King Henry II developed a single system of justice for the entire country that would be under the control of the king. Judges play an important role in making the law in common law jurisdictions in that their decisions establish legal principles, and the system operates on the principle of *binding precedent*, where the judge in a particular case must follow the decision of judges in previous similar cases.

Legal aspects of computing are concerned with the application of the legal system to the computing field. This includes the protection of intellectual property such as patents, copyright, trademarks and trade secrets, and the resolution of disputes that arise between two or more parties. Patents provide legal protection for intellectual ideas such as an original invention that is more than the next step

1.6 Legal Aspects of Computing

Table 1.3 Some areas of law in computing

Type	Description
Intellectual property	This area of law protects the intellectual property of a computer company, and may include protection of patents, copyright and trademarks
Licensing of software	Software is protected by copyright law, which helps to prevent unauthorised copies of the software being made. The licensing of proprietary software provides additional protection to the vendor
Bespoke software development	This involves outsourcing the development or testing of the software to another company, and involves a legal agreement (contract) between the two companies that defines the deliverables to be produced, the timelines, the responsibilities of both parties, and so on
Electronic commerce	This includes the legal framework for transactions to place an order, the acknowledgement of the order, the acceptance of the order, the legal contract between both parties, and order fulfilment
Criminal law	This is where an individual or company is prosecuted for a criminal offence such as placing a virus on a network or gross negligence that led to damage or loss of life
Civil lawsuits	This refers to lawsuits where one party takes legal action against another party for breach of contract, or for a tort that caused loss to the plaintiff

from existing technology. Copyright law protects the expression of an idea such as the software code and trademarks provide legal protection of names or symbols (e.g., the protection of names such as "Apple" or "Amazon"). Several areas where the legal system is applicable to computing include (Table 1.3).

There are potential legal consequences to an organisation that has developed software that has serious quality problems causing harm to others, and where there is evidence that the software has been inadequately developed and tested, or that it can be shown that the development and testing practices employed are inadequate or negligent.

Software is generally subject to a license, where a software license is a legal agreement between the copyright owner and the licensee that governs the use or distribution of software to the user. The two most common categories of software licenses that may be granted under copyright law are those for *proprietary software* and those for *free open source software*. The software license agreement generally provides limited warranties on the quality of the software, and limited remedies where the software is defective. Intellectual property is described in more detail in Chap. 10.

There are potential legal implications on both parties during bespoke software development and test outsourcing, where a legal contract is prepared between the supplier and the customer. This will generally include a statement of work that stipulates the deliverables to be produced, and it may also include a service level agreement and an Escrow agreement. Such contracts specify what will be delivered and when as well as quality expectations, and milestone payments are generally linked to delivery and acceptance of the agreed deliverables at key milestones. Such agreements often provide greater legal remedy than software that has been provided under license, as there is a clear contract between both parties with obligations on both parties. Software outsourcing is described in more detail in Chap. 9.

Computer crime includes the unauthorised access of computer resources, the theft of personal information, cyber extortion, and denial of service attacks. The problem of hacking is where a hacker uses his (or her) computer skills to gain unauthorised access to a computer system. We distinguish between ethical white hat hackers and malicious black hat hackers, where white hat hackers play a role in improving system security, whereas black hat hackers seek to exploit vulnerabilities for financial or malicious gain. Computer crime is described in more detail in Chap. 12.

Electronic commerce includes transactions to place an order, the acknowledgement of the order, the acceptance of the order where a legal contract now exists between both parties, and order fulfilment. The contract specifies the terms and places responsibilities on both parties, and such contracts generally have a cooling off period, where the buyer may cancel the contract without penalty (but the buyer would generally be subject to the costs involved in returning the goods in the case of cancellation). Legal and ethical aspects of electronic commerce are described in Chap. 11.

1.6.1 Intellectual Property

Intellectual property law deals with the rules that apply in protecting inventions, designs and artistic work, and in enforcing such rights. The inventor is generally granted exclusive rights to the invention for a defined period, and this provides an incentive to inventors to develop creative works that may benefit society as the owner of the invention is granted exclusive rights to profit from their work.

The main forms of intellectual property are patents, copyright and trademarks. Patents give inventors exclusive rights to their invention for a specified period (possibly up to 20 years). It protects innovative ideas and concepts, and the invention itself must be novel and more than an obvious next step from existing technology. The patent gives the inventor protection against patent infringement in a specific country or region of the world.

A *copyright* applies to original writing, music, motion pictures and other original intellectual and artistic expressions. It does not protect the underlying idea as such, and what is protected is the expression of the idea. Copyrights are exclusive rights to making copies of the expression, where the ways of expressing ideas is copyrightable. Copyright law protects computer software source code from being copied by third parties without obtaining the required permission. The term *"fair use"* refers to the permitted limited use of copyrightable material without acquiring permission from the copyright owner.

A *trademark* protects names or symbols that are used to identify goods or services, and their purpose is to avoid confusion and to help customers to distinguish one brand from another. Legal aspects of intellectual property are discussed in more detail in Chap. 10.

1.6.2 Software Licenses

Software developers and testers often employ dedicated tools for various parts of the software development and testing process, and the use of tools is generally subject to a licensing agreement. The tools may be developed in-house, but it is more common to employ proprietary tools or open-source tools. A software license is a legal agreement between the copyright owner and the licensee, which governs the use or distribution of software to the user (licensee). The license may cover the entire site, several users, or a single user. The organisation must satisfy the licensing agreement, and must have sufficient licenses for the deployed version of the software on site.

Computer software code is protected under copyright law in most countries, and a typical software license grants the user permission to make one or more copies of the software, where the copyright owner retains exclusive rights to the software under copyright law. The two most common categories of software licenses that may be granted under copyright law are those for *proprietary software*, and those for *free open source software* (FOSS). The rights granted to the licensee are quite different for each of these categories, where the user has the right to copy, modify and distribute (under the same license) software that has been supplied under an open-source license, whereas proprietary software typically does not grant these rights to the user.

The *licensing of proprietary software* typically gives the owner of a copy of the software the right to use it (including the rights to make copies for archival purposes). The software may be accompanied with an end-user license agreement (EULA) that may place further restrictions on the rights of the user. There may be restrictions on the ownership of the copies made, and on the number of installations allowed under the term of the distribution. The ownership of the copy of the software often remains with the copyright owner, and the end user must accept the license agreement to use the software.

The most common licensing model is per single user, and the customer may purchase a certain number of licenses over a fixed period. Another model employed is the license per server model (for a site license), or a license per dongle model, which allows the owner of the dongle use the software on any computer. A license may be perpetual (it lasts forever), or it may be for a fixed period (typically one year).

The software license often includes maintenance for a period (typically one year), and the maintenance agreement generally includes updates to the software during that time and it may also cover a limited amount of technical support. The two parties may sign a service level agreement (SLA), which stipulates the service that will be provided by the service provider. This will generally include timelines for the resolution of serious problems, as well as financial penalties that will be applied where the customer service performance does not meet the levels defined in the SLA.

Free and open-source licenses are often divided into two categories depending on the rights to be granted in distribution of the modified software. The first category aims to give users unlimited freedom to use, study and modify the software, and if the user adheres to the terms of an open source license such as the Free Software Foundation (FSF) GNU or General Public License (GPL), the freedom to distribute the software and any changes made to it. The second category of open source licenses give the user permission to use, study and modify the software, but not the right to distribute it freely under an open source license (it could be distributed as part of a proprietary software license).

1.6.2.1 Software Licenses and Failure

Software license agreements generally include limited warranties on the quality of the licensed software, and they often provide limited remedies to the customer when the software is defective. The software vendor typically promises that the software will conform to the software documentation for a specified period of time (the warranty period), and the software warranty generally excludes problems that are not caused by the software, or problems that are beyond the software vendor's control.

The customers are generally provided with limited remedies in the case of defective software. For example, the remedy provided may be an offer to replace the defective software with a corrected version, or termination of the user's right to use the defective software and a partial refund of the license fee. There is generally no financial compensation for loss or damage, and this is generally excluded in the software licensing agreement.

Software licensing agreements are generally accompanied by a comprehensive disclaimer that protects the software vendor from any liability (however remote) that might result from the use of the software. It may include statements such as *"the software is provided 'as is', and that the customers use the software at their own risk"*.

A limited warranty and disclaimer limits the customer's rights and remedies if the licensed software is defective, and so the customer may need to consider how best to manage the associated risks. Table 1.4 discusses various lawsuits that could potentially be launched against a software provider for defective software that resulted in loss or damage to the plaintiff.

1.6 Legal Aspects of Computing

Table 1.4 Types of Lawsuits

Type	Description
Criminal	This type of lawsuit is brought by the state against the software company or individuals (e.g., developers or testers) for committing a criminal act (e.g., tampering with a computer or loading a virus onto a computer)
Tort	This type of lawsuit is brought by an individual(s) against a company or individual(s) (e.g., developers or testers) for committing some wrong to him or his computer (e.g., releasing a virus onto his computer)
Negligence	The company has a duty of care to take all reasonable measures to make the product safe to ensure that the public does not suffer personal injuries or damage or loss of their property. The company could be judged to be negligence if it employed inadequate software development and testing practices, or if the staff did not have the right expertise to develop/test the software
Malpractice	This is where the quality of service is judged against a professional standard and deemed to be negligent, with mistakes made in the delivery of the service that would not be made by an ordinary professional in the field
Strict liability	A product defect caused a personal injury or damage to property, and the burden of proof required is to demonstrate that the program was defective and that the defect caused the accident (e.g., the failure of the program controlling the brakes in the car led to the car crash)
Fraud	The company made a statement of fact to you when it knew that the statement was false, and where you relied on the statement to make an economic decision such as buying a defective product
Regulatory	The regulatory sector (e.g., FDA) places requirements on how software should be developed and tested so that it is safe for the public to use. The plaintiff is required to prove that the defendant breached the defined regulations leading to loss/injury to the plaintiff
Breach of contract	A software contract specifies the obligations that both parties have to each other (as well as implied terms such as implied warranty). A breach occurs when one party fails to honour their contractual obligations

1.6.3 Legal Aspects of Outsourcing

The bespoke development and/or testing of software have become popular in the software engineering field. This may involve the outsourcing of the complete project (including the development and testing), or perhaps just the outsourcing of the software testing to an independent external organization. Bespoke (or custom) software is software that is developed for a specific customer or organization, and it needs to satisfy the defined customer requirements.

The organization will need to be rigorous in its selection of the appropriate supplier, as it is essential that the selected supplier has the appropriate knowledge and experience, and is capable of delivering high-quality and reliable software solution on time and on budget. The contract should not be awarded on costs alone, as this is just one criterion among several other important criteria.

This means that the capability of each candidate supplier as well as the associated risks needs to be clearly known prior to selection. The selection is based on objective criteria such as cost, previous working experience (if any) with the supplier, the planned approach to develop the solution, the ability of the supplier to deliver the required solution, and the supplier capability. Although, cost is an important factor in the selection, it is just one among several other important factors to consider. Often, weightings will be employed to reflect the relative importance of the criteria.

Once the selection of the supplier is finalised a legal agreement is drawn up between the contractor and supplier, which states the terms and conditions of the contract, as well as the statement of work. The *statement of work* (SOW) details the work to be carried out, the deliverables to be produced, when they will be produced, the personnel involved, their roles and responsibilities, any training to be provided, and the standards to be followed. The agreement will need to be signed by both parties, and may (depending on the type of agreement) include a warranty period, a service level agreement, training, user guides and manuals, customer support, and an escrow agreement (Fig. 1.2).

Sometimes, it will be just the testing part of a project that is outsourced, and so this is concerned with the selection and management of an appropriate supplier to perform the testing. It is essential that the selected test organisation is capable of

Fig. 1.2 Legal contract. Creative Commons

carrying out the required testing to the defined quality standard, as well as being capable of completing the testing within the budget and schedule constraints.

The legal contract defines the obligations on both the supplier and the organisation, and should either fail to honour their commitments they may well be in breach of contract. For example, the contract/SOW places obligations on the supplier to deliver various deliverables at defined dates during the project, and these deliverables need to be delivered on time as well as satisfying defined quality standards. Further, the contract will detail milestone payments to be made by the organisation to the supplier provided the agreed deliverables have been completed to the right quality standard at the committed date.

It may be that one or more parties does not honour their agreement or there may be a dispute as to whether what is defined in the contract has been honoured or not. The contracting organisation may claim that the binding agreement has not been honoured, and there may be a need to seek legal remedy if a *material* breach of the contract has occurred. However, the supplier may counter-claim that the contracting organisation is in breach of contract for failing to make the specified milestone payments.

The first step is dialogue between both parties with the objective of finding a reasonable resolution, but if both parties are unable to agree a way forward the first party may seek a legal remedy in a civil court. The legal and ethical aspects of software outsourcing are described in Chap. 9.

1.6.4 Tort in Software Engineering

The *law of tort* refers to a civil wrong where one party (the *defendant*) is held accountable for their actions (by the *plaintiff*). There are several actions that the defendant could be held accountable: e.g., negligence, trespass, misstatement, product liability, defamation, and so on. For example, the defendant may be accused of negligence and a breach of his duty of care, where damage that was reasonably foreseeable was caused by his negligence.

The impact of a flaw in software may be catastrophic, and so a software development organization must take all reasonable precautions to prevent the occurrence of defects (as otherwise it could be sued for negligence). This is especially true in the safety critical domain, where failure could cause major damage or even loss of life among the public. Therefore, software development organisations must take reasonable precautions such as having appropriate software engineering practices in place to allow the organization to consistently produce high quality software, and for stringent processes to validate the requirements and to verify that the implementation satisfies the specification.

A quality management system indicates that the organization takes software quality seriously, and that it has a sound software development process in place that serves the needs of the organization and its customers. Modern quality assurance systems include processes for software inspections, testing, quality audits, customer satisfaction, software development, project planning, etc.

The organization will need to provide evidence that it took all reasonable steps in the design, development and testing of the software to ensure that a quality product was produced. This generally includes records of the various quality assurance activities that took place during the project, and showing that there is a sound quality management system in place, and that it is appropriate and fully operational within the organization.

Another words, it is important to maintain records and an audit trail of the various quality assurance activities for a period of time after the project, so that it may defend itself should a customer decide to take legal action against it for negligence following a serious problem with the software at the customer site. The records will allow the organization to prepare a legal defence to show that it took all reasonable steps during the software development and testing, and that its behaviour was professional and fit for purpose at all times.

That is, the presence of records may be used to demonstrate that all reasonable steps were taken during the software development and testing, and the records may include lists of all the deliverables in the project; minutes of project meetings; records of risk and issue logs, records of reviews of requirements, design, and software code, records of test plans and test results; and so on.

1.6.5 Lawsuits in Computing

A lawsuit is a proceeding by one party (or several parties) against another party (or parties) in a civil court. The basic principles of litigation are where the plaintiff sues another person(s) for being negligent, and the negligence of the defendant caused injury or damage to the property of the plaintiff, and the plaintiff is seeking compensation for her loss. It involves proving in a court of law that:

The defendant had a duty of care.
The defendant breached this duty of care.
The breach caused harm to the plaintiff or to the property of the plaintiff.

The plaintiff is entitled to compensation up to the full value of the injury or up to the full value of the damage to the property if the case is successfully proved. Further, if there is clear evidence that the defendant acted maliciously or fraudulently then punitive damages may be awarded to the plaintiff to punish the defendant. Punitive damages are generally awarded in a small percentage of lawsuits, and they may be appealed to a higher court.

There are several types of lawsuit that may be brought against a software company (the defendant) including (Table 1.4).

1.7 Computer Crime

Computer crime (or *cybercrime*) is a crime that involves a computer and a network. The computer may be the vehicle by which the crime was conducted, or it may be the target of the crime. Today, more and more individuals and companies are on line, and networking systems and computers have become quite complex. There has been a major growth in attacks on businesses and individuals, and so it is essential to consider computer and network security.

The introduction of the World Wide Web in the early 1990s transformed the world of computing, and it later led to an explosive growth in attacks on computers and systems. The Internet was developed based on trust with security features added later as a response to various types of attacks, as hackers and malicious software sought to exploit known security vulnerabilities. It is therefore essential to develop secure systems that can deal with and recover from such external attacks (Fig. 1.3).

One of the earliest Internet attacks was back in 1988 when a graduate student from Cornell University released a program on the Internet (an Internet Worm) that exploited security vulnerability in the mail software to automatically replicate itself locally and on remote machines. It affected lots of machines and effectively shut down the Internet for 1–2 days.

Humans face danger on some streets or neighbourhoods in urban areas, and such dangers need to be managed. Similarly, the Internet has dangers with hackers, scammers, and web predators lurking in the shadows, and ready to pounce on those who are not well prepared or defended. There are several threats associated with network connectivity such as *unauthorised access* (a break-in by an unauthorised person), *disclosure of sensitive information* to people who should not have access

Fig. 1.3 Dandy pickpockets (1818)

to the information, and *denial of service* (DoS), where there is a degradation of service that makes it impossible to access the web site and perform productive work. Some of the threats facing a user include:

Unauthorised access
Disclosure of sensitive information
Denial of service
Theft of credit card data
Bank fraud
Defacement of web sites
Phishing emails
Virus
Cyberextortion
Ransomware
Various internet scams.

A hacker may be accessing a computer resource without authorisation with the intention of committing an unlawful act. The hacker's activities may be limited to *eavesdropping* (listening to a conversation), or it may be an active *man-in-the-middle* attack, where the hacker possibly alters the conversation between the two parties.

There may be attacks that lead to defacement of the web sites; bank fraud; theft of credit card numbers; hoax (scam) letters; phishing emails appear to come from legitimate parties, but contain links to a site that is different from the one that the user expects to go to; intercepting of packets and password sniffing. *Phishing* is an attempt by the attacker to obtain sensitive personal information such as usernames, passwords and credit card details with the intention of committing fraud.

A computer *virus* is a self-replicating computer program that is installed on the user's computer without consent. It is malicious software in that when it is executed it replicates itself and infects other computer programs by modifying them. A virus often performs some type of harmful activity on the infected computers such as accessing private information, spamming email contacts or corrupting data. It is not a crime per se to write a computer virus or malicious software. However, if that software or other malware spreads to other computers, then it could be considered a crime.

Cyberextortion is a crime that involves an attack, or threat of an attack, accompanied by a demand for money to stop the attack. They are often initiated through malware in an email attachment. These may include denial of service attacks or *ransomware* attacks that encrypts the victim's data. The victim is then offered the private key to resolve the encryption in return for payment. Companies need to manage the risks associated with cyberextortion, and to ensure that end users are properly educated on malware and phishing, and exercise good cyber hygiene to improve online security.

Another form of computer crime is Internet fraud where one party is intent on deceiving another. Among these are hoax email scams, which are designed to deceive, and fraud the email recipient. These may include the *Nigeria 419* scams, where the email recipient is offered a share of a large amount of money trapped in their country, if the recipient will help in getting the money out of the country. The recipient may be asked for their bank account details to help them to transfer the money (this information will later be used by them to steal funds), or the request may be to pay fees or taxes to release payment with further fees requested. Of course, the money will never arrive, and if a recipient receives a message that sounds too good to be true then in all probability it is a scam.

The unauthorized access of a computer system and the theft of confidential data and disruption of its services is unlawful, and may be subject to prosecution and the full rigour of the law. Computer crime is discussed in more detail in Chap. 12.

1.7.1 Hacking

A *hacker* is a person who uses his (or her) computer skills to gain unauthorised access to computer files or networks. A hacker may break into systems and cause damage or steal confidential information. *Ethical hackers* are former hackers who play an important role in the security industry in testing network security, and in helping to create secure products and services. Malicious hackers (also called *crackers*) are generally motivated by personal gain, and they exploit security and system vulnerabilities to steal, exploit or sell data.

Many computer systems in use today have vulnerabilities that may be exploited by a determined hacker to gain unauthorised entry to the system, and access to unauthorised information. It is essential to develop safe and secure systems, and known vulnerabilities in system security need to be resolved promptly by updates to the system software. Further, it is essential to educate staff on security (staff are often the weakest link), and to define (and follow) the appropriate procedures to prevent security breaches.

The security of the system refers to its ability to protect itself from accidental or deliberate external attacks, which are common today since most computers are networked and connected to the Internet. There are various security threats in any networked system including threats to the confidentiality and integrity of the system and its data, and threats to the availability of the system.

Therefore, controls are required to enhance security and to ensure that attacks are unsuccessful. Encryption is one way to reduce system vulnerability, as encrypted data is unreadable to the attacker. There may be controls that detect and repel attacks, and these controls are used to monitor the system and to take appropriate action to shut down parts of the system or restrict access in the event of an attack. There may be controls that limit exposure (e.g., insurance policies and automated backup strategies) that allow recovery from the problems introduced. Computer security is discussed in more detail in Chap. 12.

1.8 Review Questions

1. What is intellectual property law?
2. Describe the behaviours of the ethical software tester.
3. How can a software company demonstrate that it took all reasonable steps to deliver a high-quality software product, and that the testing was fit for purpose.
4. Explain the different types of software licensing.
5. Explain the legal aspects of bespoke software development.
6. What happens when one party in a test-outsourcing project believes that a material breach of the contract has occurred?
7. What types of lawsuits could be brought against a software company?
8. Explain the difference between ethical and malicious hackers.
9. What is the difference between criminal law and civil law.

1.9 Summary

Business ethics are concerned with ethical principles and problems that arise in a business environment. They refer to the core principles and values of the organization, and guide employees in carrying out their work, and include the rights and duties of a company, its employees, customers and suppliers.

Legal aspects of computing are concerned with the application of the legal system to the computing field. It includes intellectual property law including patents, copyright, trademarks and trade secrets; bespoke software development; test outsourcing; licensing of software; professional negligence in the development and testing of software; and computer crime.

A lawsuit is a proceeding by a party against another party in a civil court where the plaintiff sues another person for being negligent, and the negligence of the defendant caused injury or damage to the property of the plaintiff.

Bespoke software (or custom software) is software that is developed for a specific customer or organization, and needs to satisfy specific customer requirements. The legal contract specifies the obligations of the supplier, and should the supplier fail to honour its commitments it may well be in breach of contract. This may result in the first party seeking a legal remedy in a civil court.

A software license is a legal agreement between the copyright owner and the licensee, which governs the use or distribution of software to the user (licensee). Computer software code is protected under copyright law, and the license grants the user permission to make one or more copies of the software. Software license agreements generally provide limited remedies to the customer when the software defective. However, there may be legal implications if the software has been inadequately developed and tested.

A hacker is a person who uses his (or her) computer skills to gain unauthorised access to computer files or networks. Hackers may probe parts of the system for weaknesses, and system vulnerabilities may lead to attackers gaining unauthorized access to the system. The system needs to be designed for security, as it is difficult to add security after the system has been implemented.

References

1. B.J. Lonergan, *Insight. A Study of Human Understanding* (Harper Collins, New York, 1978)
2. R.C. Barquin, *In Pursuit of a 'Ten Commandments' for Computer Ethics* (Computer Ethics Institute, 1992)

A Short History of Ethics

2

Abstract

This chapter presents a history of ethics, and we discuss ethics in ancient civilisations as well as ethics in several religious traditions. We discuss utilitarian and deontological ethics and consider the question of what it means to be an ethical person in a complex world.

Keywords

Hammurabi code • Ancient Greek ethics • Virtue ethics • Relativism • Christian ethics • Islamic ethics • Buddhist ethics • Chinese ethics • Utilitarianism • Deontological ethics

2.1 Introduction

Ethics is a practical branch of philosophy that deals with moral questions such as what is the right thing to do in a particular situation, and in dealing with conflicts between individual moral values and the values of an organisation or society. How should one live a good life? How should a person behave in a given situation in a complex world? Ethics explores what actions are right or wrong within a specific context or within a particular society, and attempts to find satisfactory answers to moral questions. The origin of the word "ethics" is from the Greek word ἠθικός, which means habit or custom. Ethics provides a framework to think critically about moral problems and moral values, and assist in forming a judgment on the most appropriate way to proceed in a situations where there is a conflict between values, and where the individual is trying to do the right thing.

There are various schools of ethics such as the *relativist* position (as defined by Protagoras and described in Plato's Theaetetus), which argues that each person decides on what is right or wrong and so it rejects the idea of objective truth; *cultural relativism* (which is a variant of the relativist position) argues that the particular society determines what is right or wrong based upon its cultural values; *deontological ethics* (as defined by Kant) argues that there are moral laws to guide people in deciding what is right or wrong, and that everyone has a duty to act in accordance with the moral law; and utilitarianism (as defined by Bentham and refined by Mill) which argues that an action is right if its overall affect is to produce more happiness than unhappiness in society.

Religion plays an important role in developing ethical and moral behaviour in society, and we discuss the contribution of a selection of the great religions of the world including Christianity, Islam and Buddhism. We briefly discuss traditional Chinese ethics; utilitarianism as proposed by Bentham and Mill; deontological ethics as proposed by Kant, and virtue ethics as described in ancient Greek philosophy. Finally, we discuss informally how an individual can be ethical in a complex world, and how to create an ethical environment in an organisation to support ethical decision-making. A short history of ethics is available in [1].

2.2 Ethics in Ancient Civilisations

The changes in human lifestyle from traditional nomadic or migratory societies to settled communities posed new challenges as to how these communities should live together in peaceful harmony, and to how they should resolve any disputes that arose between them. This led to basic regulation of settled communities to protect people and property, and often there was a role for the elders in the community to mediate and resolve conflicts that arose between members of the community. Over time, this led to a systematic legal code to regulate behaviour, where victims were compensated for damage to property or the person, and offenders were punished for their misdemeanours.

One of the earliest systematic codes was the famous Hammurabi code which was introduced in Mesopotamia[1] c.1750 B.C., and it is named after King Hammurabi of Babylon. The code focused on awarding harsh penalties to the perpetrators of crime, and an appropriate physical punishment was given for specific crimes. It is one of the earliest legal codes where it is presumed that the accused person is innocent until proven guilty (Fig. 2.1).

Personal injury involving the aristocracy involved the principle of *lex talionis* (the principle of retribution: i.e., an eye for an eye), where the perpetrator received

---

[1] Mesopotamia was located in modern Iraq, Syria and Turkey between the Tigris and Euphrates rivers, and it is known as the "cradle of civilisation". It consisted of several civilisations spanning a period of roughly 3000 years, which included the Sumerians, the Akkadians, the Babylonians, the Hittites and the Assyrians. Some of the famous Mesopotamian sites include Ur, Babylon, Nimrud, and Nineveh. Mesopotamia led to many important inventions that changed the world such as the wheel, maps, and writing. Mesopotamia came under the control of Cyrus the Great of Persia c. 539 B.C., and Alexander the Great conquered the Persian Empire c. 331 B.C.

Fig. 2.1 Code of Hammurabi Stele. Louvre Museum. Creative Commons

as punishment precisely those injuries or damage that they had inflicted on their victims, whereas fines sufficed for injuries to slaves or freemen. That is, the law protected the aristocrat over the freeman and slave.

Careless or inefficient behaviour could result in fines for damages, with serious crimes subject to harsh punishments such as the removal of the perpetrator's tongue, hands, eye or ear. There are over 280 edicts in the code that are written in the form of IF–THEN statements covering family law, professional contracts and administrative law, with different standards of justice provided for the three classes in Babylonian society (i.e., the nobility, free people not belonging to the upper class, and slaves).

The ethical system at the heart of the ancient Egypt was based on *ma'at*, a word that signified justice or balance, and it provided the foundation for ethical behaviour and judgment. Egyptians were expected to live in harmony with *ma'at*, and ethical norms that sustained *ma'at* were taught in scribal schools. The Egyptians had a strong belief in the afterlife and had a concept of ethical judgement of the deceased person.[2] The Egyptian 'Book of the Dead' listed sins that the deceased would recite and give a negative confession.

---

[2] A part of this ritual involved weighing the deceased person's heart against a feather, and if the heart weighed less than a feather (i.e., light hearted) then the deceased was judged to be a good person.

2.3 Ethics in Ancient Greece

The Greeks made major contributions to western civilization including mathematics, logic, astronomy, philosophy, politics, drama, and architecture. The Greek world of 500 B.C. consisted of several independent city-states such as Athens and Sparta in what is now mainland Greece, and various city-states in Asia Minor such as Ephesus, Miletus and Pergamon. The Greek polis (πολις) or city-state tended to be quite small, and it consisted of the Greek city and a certain amount of territory outside the city. Each city-state had its own unique political structure for its citizens: some were oligarchs where political power was maintained in the hands of a few individuals or aristocratic families; others were ruled by tyrants (or sole rulers) who sometimes took power by force, but who often had a lot of support from the public. These included people such as Solon, Peisistratus and Cleisthenes in Athens.

The reforms by Cleisthenes in the sixth century B.C. led to the introduction of the Athenian democracy. Power was placed in the hands of the citizens who were male (women or slaves did not participate in Athenian democracy). It was an extremely liberal democracy where citizens voted on all of the important issues. Often, this led to disastrous results as speakers who were skilled in rhetoric could exert significant influence in the Athenian assembly (e.g., the decision that the assembly made that led to the disastrous Sicilian expedition during the Peloponnesian war). The failures of Athenian democracy (including its decision to execute the well-known moral philosopher, Socrates) led Plato to reject democracy as a system of rule, and instead he advocated rule by philosopher kings as described in Plato's Republic.

The rise of Macedonia led to the Greek city-states being conquered by Philip of Macedonia in the fourth century B.C. His son, Alexander the Great, defeated the Persian Empire, and he extended his empire to include most of the s known world. This led to the *Hellenistic Age* where Greek language and culture spread throughout the empire. Alexander founded the city of Alexandria, and it became a major centre of learning in Ptolemaic Egypt.[3] However, Alexander's reign was very short as he died at the young age of 33 in 323 B.C

Ethics in Greek philosophy is concerned with rational thought and the search for appropriate principles to guide human conduct. It is characterised by its faith in reason to do the right thing. The fundamental question asked by Socrates and other Greek philosophers was how should a man[4] live, in order to achieve a good life (*eudaimonia*)? That is, how should a man live in order that we may reasonably say that he has lived successfully? Another words, what is the nature of the good life for man? Socrates argued that the ways of man can be and need to be justified by rational means, and ultimately in terms of an individual's self interest, and *arête* (moral virtue or excellence) is necessary for a good life.

---

[3] The ancient library in Alexandria was once the largest library in the world. It was build during the Hellenistic period in the third century B.C. and destroyed by fire in 391 A.D.

[4] Women or slaves did not participate in the Athenian democracy!

Socrates[5] was a moral philosopher who actually wrote nothing on philosophy, but a clear picture on his views and approach to philosophy may be seen in, Plato's early dialogues.[6] Plato's writings are in dialogues involving his former mentor Socrates, with the early dialogues reflecting the views of the historical Socrates, with the later dialogues using Socrates as a mouthpiece to reflect the reviews of the mature Plato.

Socrates claimed a total lack of knowledge, and he often engaged in dialogue with the citizens of Athens with the goal of clarifying and clearly defining key concepts. Often, his interaction with the citizens of Athens demonstrated that those considered wise were really not that wise after all, and knew even less about the object of enquiry than he did,[7] and he himself claimed no knowledge whatsoever. Another words, his approach of forensically examining his opponent's argument and showing that it led to contradictions or nonsense had the effect of making his opponents look silly, and hardly made him popular in Athenian society.

Socrates was also damaged by his association with Alcibiades, who was a flamboyent Athenian statesman and orator. Alcibiades was a strong proponent of the disastrous Sicilian expedition during the Peloponnesian war, and he also changed his allegiance several times during the war (he also served as a strategic advisor to Sparta for a short time).

Socrates answer as to how man should live a good life is defined in terms of *arete* (which is generally translated as moral virtue), and where a good life is considered to be a good moral life. The right thing to do, according to Socrates, involves thinking hard about the situation, and *doing the right thing is best for you.* Further, if anyone does wrong it is because they have not thought enough about the situation, since, according to Socrates, no one does wrong willingly. Socrates claims not to know what virtue is, and he explores and questions others in an attempt to define and give an account of *arete* in the early Platonic dialogues.

Aristotle was born in Northern Greece and he became a student of Plato at Plato's academy in Athens in the fourth century B.C. (Fig. 2.2). Aristotle later founded his own school (known as the Lyceum) in Athens, and he also became

---

[5] Socrates was a moral philosopher who deeply influenced Plato. His method of enquiry into philosophical problems and ethics was by questioning the citizens of Athens. Socrates himself maintained that he knew nothing (Socratic ignorance). However, from his questioning it became apparent that those who thought they were clever were not really that clever after all. His approach obviously would not have made him very popular with the citizens of Athens. Socrates was sentenced to death for allegedly corrupting the youth of Athens, and he was forced drink the juice of the hemlock plant (a type of poison).

[6] Socrates did not write any works on philosophy, and knowledge of his approach to moral philosophy is described in Plato's early dialogues, whereas the "Socrates" of the later dialogues reflects Plato's mature philosophical thinking rather than that of the historical Socrates.

[7] Chaerephon (a lifelong friend of Socrates) asked the oracle at Delphi who was the wisest of all men, and the oracle replied that there was none wiser than Socrates. Socrates set out to determine what the oracle meant by this statement, and in his discussions with the wise of Athens he realised that those considered wise really knew nothing but thought that they did, whereas he alone was aware that he knew nothing (Socratic Ignorance).

the tutor of Alexander the Great. He made contributions to biology, logic, politics, ethics and metaphysics. His starting point to knowledge acquisition was the senses, as he believed that these were essential to acquire knowledge. His position is the opposite of Plato who argued that the senses deceive and should not be relied upon, and that knowledge is more than just sense perception. Instead, for Plato knowledge is eternal and unchanging, and consists of truths that are not objects of perception.

Aristotle wrote the Nicomachean Ethics and the Eudemian Ethics, and his position was that people should have an excellent character (virtuous character) in order to achieve happiness (or well being). He argues that reason distinguishes man from animals, and that the human function is some kind of excellent exercise of the intellect. His approach differs from Plato's in that his analysis of ethics uses his metaphysical theory of potentiality and actuality, with happiness defined as an actuality.

Aristotle argues that moral virtue or excellence of character may be taught in a similar way to learning a musical instrument: i.e., ethical character is a skill learned through practice, which a person develops partly as a result of upbringing and partly as a result of habit of action. That is, for Aristotle the source of ethical

Fig. 2.2 Plato and Aristotle

insight is in life experience, and so knowing how to act is based on practical wisdom developed from a combination of training in the right habits and experience with practical situations.

Aristotle formulated the doctrine of the *'Golden Mean'*, which locates each of the virtues between corresponding vices of excess and defect. He defines the four cardinal virtues in a similar way to Plato, and they include prudence, temperance, courage and justice. He describes *eudaimonia* as the highest human good with *arête* as the highest sense achieved by the life of reason and the intellect.

Aristotle made important contributions to formal reasoning with his development of syllogistic logic. Syllogistic logic (also known as term logic) consists of reasoning with two premises and one conclusion. Each premise consists of two terms and there is a common middle term. The conclusion links the two unrelated terms from the premises.

Stoicism was founded by Zeno of Citium, Cyprus, in the third century B.C., and remained popular in the Greco/Roman world until the third century A.D. The Stoics taught that virtue is essential for human happiness, and Stoics are often seen as being unmoved by the sorrows and afflictions affecting humanity. They aimed in a sense to live for the moment and to rationally understand the world, and to avoid being controlled by emotions and passions, as they rejected these as a way to decide what is good and bad.

Eudaimonia is generally translated as happiness or well being by the Greeks, but it was identified as pleasure by Epicurus who was the founder of the Epicurean school of ethics/philosophy in the late fourth century B.C. Epicurus was a hedonist, who believed that pleasure is morally good and pain is morally evil, and that the absence of pain or fear is the greatest pleasure.

It should be noted that the institution of slavery was taken for granted in ancient Athens, and that women played a subordinate role in role in Greek society. The Greek philosophers remained in a sense, prisoners of their own culture, and took these prejudices and inequalities for granted.

2.4 Ethics in Religious Traditions

The great religious of the world provide a moral compass to their communities to behave ethically in the world. We discuss Christianity, which arose as a reform movement of Judaism in Palestine during the first century A.D.; Islam, which was founded by the prophet Mohammed in the Arabian Peninsula during the seventh century A.D.; and Buddhism, which was founded by the Buddha, Siddhãrtha Gautama, in Northern India c. 500 B.C.

2.4.1 Christian Ethics

The Christian faith is deeply ethical and its sources are in the Bible and in the life of Jesus. The bible consists of the Old Testament and the New Testament, with the New Testament focusing on the life of Jesus and his death and resurrection. The books of the Old Testament are based on the twenty-four books of the Hebrew Bible (*Tanakh*), which record the history and life of the Jewish people, their covenant with God, their laws, their kings, the destruction of Jerusalem and their exile to Babylon, their return to Zion, their belief in the coming Messiah, and the lives of the prophets.

The Old Testament includes books such as Genesis, Exodus, Leviticus, and Deuteronomy; the sayings of prophets such as Joshua, Jeremiah, Isaiah, and Ezekial; the Psalms, which are songs and prayers for worship; Proverbs, which are sayings that contain wisdom; Job, which explores the nature of suffering; and books such as Ruth, Ecclesiastes, Lamentations and Esther that are associated with particular festivals.

The New Testament consists of twenty-seven books that include the Gospels, which present the life and teachings of Jesus, and the letters (epistles) that were written by early Christian leaders (e.g., St. Paul) to provide guidance to the early Church. The New Testament concludes with the book of Revelation, which presents a vision of things from creation to the end of the world. Christianity began as a reform movement of Judaism in Roman Palestine, and its early adherents were Jewish converts. Many Gentiles (non Jews) embraced Christianity, and it gradually spread throughout the Roman world and became a major religion in its own right.

St. Paul converted to Christianity, and he was one of the most important missionaries of the early church. He preached the gospels, established churches and contributed to spreading Christianity to the urban areas throughout the Roman Empire. The direct Jewish influence on Christianity ceased after the Jewish revolt against Roman rule from 66 to 73 A.D., which led to the destruction of the second temple in Jerusalem and the expulsion of many Jews from Judea.

Christianity became the official religion of the Roman Empire after the conversion of the emperor Constantine in 337 A.D.[8] Christians are monotheistic: i.e., they believe in one God who created the heavens and earth, and the Christian concept of the Trinity is that God consists of three persons: namely the father (God himself), the son (Jesus Christ) and the Holy Spirit.[9] Christians believe that God sent his only son, the Messiah, to save mankind, and that Christ was crucified on the cross to offer the forgiveness of sins, and was resurrected three days after his death before ascending to heaven. Christmas celebrates the birth of Jesus, and Easter celebrates the death and resurrection of Christ. The God of Christianity is all powerful (*omnipotent*), all knowing (*omniscient*), and is present everywhere

---

[8] Constantine had lifted the ban on Christianity in the Roman Empire in 313 A.D, and the emperor Theodosius I declared Christianity to be the official religion of the Roman Empire in 380 A.D.

[9] There were a number of schisms in the early Christian church, including Arianism which was a dispute on the nature of the Trinity, and whether the Father caused the creation of the Son and Holy Spirit.

Fig. 2.3 The crucifixion of Christ

(*omnipresent*). For Christians, God is good and God is love, and the Christian concept of love is powerful in that the love of God is viewed as his love of humanity, and also the love of Christians for each other as in loving one's neighbour as oneself (Fig. 2.3).

The starting point in Christian ethics is the Christian faith and the life of Jesus as recorded in the Gospels, as well as Jewish ethics from the Old Testament (e.g., the ten commandments). Ethical behaviour involves both the right motive and the right action, and since Christians believe that the Bible is the revealed Word of God, the Bible is employed to teach and learn the right behaviour. Reason is part of the foundation of ethics and must be employed with revelation, and St. Anselm indicated the close relationship between Christian faith and human reason with his statement "*fides quaerens intellectum*" (faith seeking understanding).

Jesus teaches the importance of the kingdom of God and love, and one should love God with all one's heart and to love one's neighbour as oneself. That is, the Christian message from the Gospels is one of peace, love and compassion, and that God loves us and expects us to love one another as well as loving God. The *Golden Rule of Christian Ethics*[10] is stated by Jesus in the Gospels of Saint Luke and Saint Matthew, and is in a sense the principle of reciprocity, and is a summary of the books of Laws and the Prophets of the Old Testament. It states, "*Do onto others as you would have them do onto you*", as may be illustrated as, for example, where one should be charitable to others such as the poor in society, as if you were in need then you would want others to help you. There is an interesting account of the philosophy of the Christian religion in [2].

---

[10] This principle is also part of Judaism and Islam.

Bertrand Russell[11] expressed doubt over the historical existence of Jesus, and he maintained that while Jesus had a very high degree of moral goodness that there were some flaws in his character. These included his belief in hell and everlasting punishment, and also his behaviour and reaction when individuals at a gathering did not listen to his preaching (he reacted with fury). Russell also questioned the morality of religion stating that it is an impediment to moral progress in the world [3].

2.4.2 Islamic Ethics

The prophet Mohammed founded Islam in the seventh century A.D., and it (along with Christianity and Judaism) is one of the great monotheistic religions of the world. God (called *Allah* in Islam) is Lord and creator and the sole master of mankind. The final judgment will be a terrible vengeance on the ungodly, and man's only hope is *Islam* (meaning submission to God). The Koran (*Qur'an*) meaning recitation is the word of Allah as revealed through the prophet Mohammed, and Allah urges reason to be exercised in understanding revelation. The other source of Islam is the recorded deeds and sayings of the prophet as recorded in the *Sunnah* (*hadith*),[12] and the message of the Koran and examples of the prophet's life acts a paradigm for ethical and moral behaviour for Muslims around the world.

The inhabitants of the city of Mecca initially rejected the teachings of Mohammed, and the prophet fled to the city of Medina.[13] Later he conquered Mecca with his Muslim army (entering the city without battle), and the concept of the jihad[14] (holy war) is associated with Islam. Mecca is the holiest city in Islam

---

[11] Bertrand Russell was an influential
British logician, mathematician and philosopher, and his grandfather was a former prime minister of Britain. He was the co-author with Alfred Whitehead of *Principia Mathematica*, which aimed to derive all of the truths of mathematics from logic. Russell was also active in political causes such as the campaign for nuclear disarmament. He was sympathetic to the plight of the Palestinian people, and he criticised Israeli actions in annexing the territory of the West Bank and Jerusalem after the 1967 war. He also opposed the Vietnam War.

[12] Some Islamic scholars question whether it is appropriate to attach the same importance to the Sunnah/Hadith as the Koran. This is since the former was written over a century after the death of the prophet and so errors could have been introduced given the elapsed time. Further, if the Sunnah/Hadith had been divinely revealed they would have been written down at the same time as the Koran was written down.

[13] The prophet is buried in the city of Medina (one of the three holiest cities of Islam), and the city is also visited as part of the annual Hajj and Umrah pilgrimages to Mecca. One of the earliest battles in Islam was fought at Mt. Uhud on the outskirts of Medina.

[14] Jihad also includes striving by peaceful as well as by armed means to achieve the goals of a Muslim community. The Koran specifies the conditions for war and peace in the case of an armed conflict.

Fig. 2.4 The Ka'baa in Mecca, Saudi Arabia

and it contains a cubical building called the *Ka'bah*,[15] which is the most sacred site in Islam. Muslims circle the *ka'bah* seven times in a counter clockwise direction during the *Hajj* or *Umrah*, and it is the direction of prayer (*Qibia*) for Muslims around the world. God is portrayed in places in the Koran as ready to forgive, and in other places he punishes. Christ is considered a prophet in the Koran, and it is emphasized repeatedly that God could not have a son (Fig. 2.4).

Muslims pray five times a day facing Mecca, and the muezzin announces the call for daily prayers (*salah*) and Friday prayers (*adhan*). All healthy Muslims are expected to do the *Hajj* (pilgrimage to Mecca) at least once in their lifetime. Fasting takes place during the ninth lunar month (*Ramadan*), and is strictly observed by all healthy Muslims. This involves absence from food and drink between sunrise and sunset.

Polygamy is allowed for Muslim men, but not for Muslim women. Furthermore, Muslim men are allowed to marry Christians or Jews, but this is forbidden to Muslim women. Women play a subservient role in Muslim society (especially in parts of the Middle East), and it is common for a veil to be worn for modesty purposes.[16] The Koran emphasises the importance of addressing social and

---

[15] It has been suggested by some scholars that in pre-Islamic Arabia that the Ka'baa was originally dedicated to Hubai, a Nabatean deity. Petra in Jordan was the northern capital of the Nabatean civilisation with Madain Salih near Al Ula in Saudi Arabia the southern capital. In early Islam, the direction of prayer (*qibia*) for Muslims was in the direction of Jerusalem, but this was later changed to the direction of the Ka'baa in Mecca.

[16] Some Muslim societies (e.g., in Saudi Arabia and other parts of the Arabian peninsula) are patriarchal, and require females to wear a loose long black robe termed an *abaya*, and some also require them to have their hair and face covered with a *hijab* and *niqab*. Other Muslim societies (e.g., Indonesia) are more relaxed in the dress code for females, and they may wear more colourful clothes. The Prophet emphasised that women should dress modestly, but he did not specify what this meant in terms of burka, abaya, hijab and niqab. Some of the practices that are unique to Islam

economic injustice, and Muslims are urged to spend their wealth on friends and relatives, orphans, the poor, and the needy (*zakat*).

Islam provides a moral compass for the Muslim community (*ummah*), and the revelation to Mohammed as recorded in the Koran provides the reference point for Muslims in distinguishing right from wrong. That is, divine command plays an essential role in establishing the moral order for Muslims, and the message from the Koran and sayings in the Sunnah led to the codification of these norms and to the concept of Islamic Law as recorded in *Shar'ia*.

Islam has been criticised as been ruthless and intolerant, and the Indian pacifist and political activist, Mahatma Gandhi, noted that while non-violence has a place in the Koran, the historical expansion of Islam through military conquest means that Muslim fighters are aggressive. Islam has been criticised for its prejudice against homosexuality, where homosexual acts between consenting adults carry the death penalty in some Islamic states.

The penalty for *apostasy* (renunciation of faith) in Islam is death, which is contrary to the United Nations Human Rights convention that recognises the right to change religion. Islamic Sharia law is often viewed as cruel, with the penalties that it applies being unsuitable to the modern world, and belonging more to the age of early Islam on the Arabian Peninsula. The perception of Islam as a violent religion has been compounded by the rise of extremist forms of Islam such as in Al Qaeda and the Islamic State, where violence including cruel and inhuman behaviour (e.g., decapitations) have taken place.

However, it is important to recognise (based on the author's experience of living in the Middle East, including time spent in Saudi Arabia and Kuwait) that the vast majority of Muslims seem to be decent and gentle human beings, and they are just trying their best to lead a good spiritual life in harmony with their faith. There are gradual changes taking place in several of the more conservative parts of the Middle East, and while gender inequality and Sharia law of traditional Muslim societies remains deeply disturbing, in time (hopefully soon), an appropriate balance between these traditional religious beliefs and modern life will be achieved.

Bertrand Russell noted that Islam has been a political religion from the very beginning, where Mohammed and the caliphs who followed him making themselves rulers of their communities with both spiritual and political authority up to the end of the first world war. Islamic philosophers such as Averroes (from Persia) and Avicenna (from Andalucia, Spain) played an important role in preserving and transmitting the philosophical works of Plato and Aristotle as well as providing commentaries on these works. For more information on Islam see [4].

---

in the Arabian peninsula seem to have their origin in the tribal practices of the Bedouin desert tribes in Arabia from which Islam emerged.

2.4.3 Buddhist Ethics

Buddhism is centred on the teachings of the Buddha, as recorded in written form by later Buddhists. It arose as a breakaway religion from Hinduism, and it is an atheist religion (i.e., a religion without God). Siddhārtha Gautama (the Buddha and founder of Buddhism) was born into the ruling family of a kingdom near Lumbini in North India in the sixth century B.C. He became aware of the challenges of the human condition early in his life, including the negative effects of the caste system, as well as sickness, anguish, suffering and death. He left the royal palace in his late 20s with the goal of becoming an ascetic so that he could investigate and find solutions to these problems.

He practiced existing techniques of meditation with several teachers, and he later developed his own approach that allowed a stage of calmness to be reached that facilitated the development of insight. This enabled him to reach enlightenment, and he spent the remainder of his life preaching and teaching on suffering, impermanence, and egolessness.

The Buddha's teachings are preserved in the *Pali canon*, and they were handed down in an oral tradition until monks wrote them down in the first century A.D. The Pali Canon consists of three baskets, including the *Vinayana* (Discipline for the monks), *Sutta* (Discourses or basic teachings of the Buddha), and *Abhidhamma* (Doctrinal Elaboration or a more detailed systematisation of the doctrine by later followers). The Vinayana emphasises that Buddhism is essentially about monastic asceticism.

At the core of the Buddha's teaching is the doctrine of the *four noble truths*, which summarises the core principles of Buddhist philosophy and the basic message of the Buddha. The Buddha gave the first Buddhist sermon at Sarnath in India, and the Buddha explained the four noble truths and the eight-fold path to end suffering (Fig. 2.5).

- Existence of suffering (*dukkha*)
- Origin / cause of suffering (*samudaya / trsna*)
- Cessation of suffering (*nirodha*)
- Eightfold path to end suffering (*magga*).

The four noble truths are outlined in the form of a medical diagnosis, where there is an illness; the causes of the illness are identified; a solution is proposed; and the steps to implement the solution are identified. The causes of suffering according to the Buddha are due to your mental attitude or mindset of attachment (*trsna*). Suffering can be eliminated by a change of mental attitude: i.e., by getting rid of the cause you get rid of the effect. Finally, there are a set of principles that you can implement in your life to get rid of your suffering, and these involve eight principles that cover the right view and wisdom, the right action and the right mental state.

That is, Buddhism aims to specify a path which if followed terminates the endless cycle of reincarnation and transmigration of souls characterized by the

Fig. 2.5 Buddha teaching Four Noble Truths

Hindu Religion. This path essentially involves self-discipline and asceticism, and the Buddha identified an *eight-fold path* to reach Nirvana. The Buddha does not specify what Nirvana is; rather, he specified how Nirvana is reached.

The earliest Buddhist tradition is called *Theravada* (also called *Himayana* or Small Vehicle), and is close to the teachings of the historical Buddha. It became popular in South East Asia (including Thailand, Sri Lanka and Burma). The other major tradition of Buddhism developed later in India (first century B.C.), and is called *Mahayana* (Great Vehicle), and it spread to China and later to Mongolia and Tibet.

Buddhism is concerned with leading a good life, practicing virtues and following meditation exercises. Meditation plays a role for being in the moment and concentrating to avoid distracting thoughts, and rational analysis and argument plays an important role in Buddhist ethics.

At the heart of Buddhism is the concept of *dukkha* which includes suffering, physical pain, sickness and essentially all the undesirable things that are part of the human condition. The Buddhist teaching on egolessness (*anatta*) and impermanence (*anicca*) involves a denial of the sense of self that humans' possess.[17]

---

[17] The normal view of self is that it is a part of you that remains constant throughout life and defines "you" as you (i.e., it is the essential core of you). Buddhists reject this view and the Buddhist position is analogous to that of an automobile that is made up of parts that were assembled in a factory; the parts interact with each other and the road; they may wear out and be replaced; and eventually the car reaches the end of its life. The car does not have a self as such: it does not have a part that remains constant throughout its life. You, according to Buddhism as a person are like the car and have no self.

The world of things is impermanent: i.e., things come into existence and go out of existence, and so everything passes in time.

Buddhist ethics (*sila*) are part of the eight-fold path, and they specify a code of conduct that involves a commitment to harmony and self-restraint, with the key principles being non-violence and to avoid causing harm. Ethics refers to the avoidance of non-virtuous deeds, and requires concentration and control of the mind and wisdom of the nature of reality.

The eight-fold path emphasises *right view*: i.e., understanding of the nature of things especially the 4 noble truths; *right intention*: i.e., avoiding attachment, hatred and harmful intent; *right speech*: i.e., avoiding lying, divisive speech and harsh speech; *right action*: i.e., avoid killing, stealing and sexual misconduct; *right livelihood*: i.e., avoid trades that harm others; *right effort*: i.e., avoiding negative states of mind; *right mindfulness*: awareness of body and thoughts; and *right concentration* for practicing meditation.

The Buddha stressed that wisdom (*prajna*) and compassion (*karunā*) go hand in hand and are essential to achieving enlightenment. Buddhist compassion includes sympathy for the pain and suffering of others. For more information on Buddhism see [5].

2.5 Traditional Chinese Ethics

Chinese ethics are concerned with practical problems, such as how one should live as well as one's duties to families, strangers and wider society. It also considers the extent to which one should be involved in reforming the social and political structures in society, and how one should behave when in a position of power. Chinese ethics integrate the personal, social and the political, and the process of behaving ethically is achieved in the context of the social relationships of family and wider society, and especially to those to whom one has responsibilities. That is, there is an emphasis on the family and social context in making the right ethical choices and taking the appropriate action.

The various Chinese approaches to ethics include Confucianism, Daoism and Chinese Buddhism. The *dao* (way) provides a guide to the public, and the compound term *dao de* (meaning ways and virtues) consists of the interaction of the social *dao* and *de* (virtue). *Dao* is public, objective guidance, whereas *de* consists of the character and disposition obtained from exposure to *dao*, and is the physical realisation of *dao* in some part of the human system (e.g., family). Both *dao* and *de* encompass more than just morality, and they are also ways of fashion and etiquette.

Daoism (or Taoism) is a Chinese philosophy that emphasises living in harmony with the Tao (Dao). Ethics in Daoism emphasise effortless action including naturalness and spontaneity, and the three treasures of compassion, frugality and humility. It dates back to the fourth century B.C. with its teachings attributed to Laozi and Zhuangzi.

Fig. 2.6 Confucius, Tang Dynasty

Confucius was born in the sixth century B.C., and Confucianism is a way of life that includes a system of thought and behaviour. It discusses ethics in terms of virtues and the corresponding ideals of a person. It believes humans are fundamentally good, teachable, improvable and perfectible through personal and community development. There are similarities between ethics in Confucianism and the Golden Rule of Christian Ethics, which may be illustrated with the Confucian belief *"What you do not wish upon yourself, extend not to others.* (Fig. 2.6).

Confucianism is popular in South East Asia including countries such as China, Hong Kong, Macau and Vietnam. The right behaviour involves more than instructions or a code, as the instructions need to be interpreted to the particular situation, and a person's duties are duties of their station towards other socially described persons or things.

That is, Confucianism deals with people in terms of their social relationship with them: e.g., such as parent, daughter, sibling, friend, neighbour, and so on.

2.6 Utilitarianism

Jeremy Bentham founded utilitarianism in the nineteenth century, and this approach to ethics involves measuring the consequences of an action against one value (such as benefit or happiness). The sole moral criterion for good or bad lies in the *utility principle*, which is the greatest happiness of the greatest number of a community. Another words, the utility principle is the only ground for action, and it provides a framework to act morally by taking actions that maximize *utility*, where the concept of utility is defined as the property of an object that provides benefit or happiness. That is, the *right action in a given situation is the action that produces the most benefit or happiness to society*, and so if a particular law or action doesn't do any good then it isn't any good.

Bentham argued that the greatest happiness may be determined quantitatively from an estimate of the expected pleasure or pain for each action. That is, utilitarianism allows (in a sense) a moral balance sheet to be drawn up with the costs and benefits of each action recorded, and it thereby allows the action with the best consequences to be chosen. The process to determine the action that leads to the greatest happiness for the greatest number of people involves a summation of the pleasure / pain for all individuals.

Bentham's motivation for the development of classical utilitarianism was his desire for a reform of the legal system and for social reform in England. His goal was that useless or corrupt laws and social practice should be changed appropriately. Bentham's analysis of the laws identified some that were bad due to their lack of utility: i.e., some laws were bad in the sense that they led to unhappiness and misery without any compensating happiness. Bentham argued that utilitarianism should become the standard to determine the right action to be taken by the government and individuals, and John Stuart Mills and others later promoted his ideas.

Utilitarianism is a member of the class of *Consequentialism Ethics* that argue that the consequences of one's action are the ultimate basis for a judgment as to whether the action is right or wrong. That is, an action in itself is neither right nor wrong: rather it is the consequence of the action that is morally relevant.

The utilitarian position may result in conflicts when the actions that promote overall happiness and good are incompatible with the individual's happiness. That is, one of the key criticisms of utilitarianism is that the individual is not protected when the utility calculation yields that the pleasure of the majority outweighs the unhappiness of a small number of people. This could result in the abuse of individuals and minorities, and John Stuart Mill introduced the *"no harm principle"*, where one is free to do as one wishes provided that no harm is done to others. However, this principle is difficult to apply in full as moral problems often include the risk of harm to others.

According to utilitarianism even the most fundamental principles (e.g., the rights specified in the UN Declaration of Human Rights) can be broken if the positive consequences outweigh the negatives (*the end justifies the means*). This has led to a variant of utilitarianism (*rule utilitarianism*) that recognises the existence of moral rules, and looks at the consequences of rules (in contrast with actions) to increase happiness.

2.7 Deontological Ethics

Deontological ethics (from the Greek δέον meaning duty, and λόγος meaning study, i.e., the study of duty) argues that the ethical value of an action should be judged on whether the action is right or wrong in itself (based on a set of rules),

Fig. 2.7 Immanuel Kant

and not on the actual outcome or consequences of the action. That is, an action is morally right if it is in agreement with a moral rule, and this is independent of the consequence of the action. A moral obligation may arise from a set of rules from a particular religion, a set of personal ethical values, cultural values, or rules inherent in the natural world.

Immanuel Kant[18] developed the most well known system of deontology, and a key concept in Kantian ethics is *autonomy*, where Kant argues that an individual should be able to determine what is morally correct through reasoning (i.e., reasoning is the source of morality). The key idea is that we place a moral norm or law upon ourselves, and we have a duty to obey and respect this moral norm, and if we do this we are acting with *good will*. Another words, an action is morally right if it is in agreement with a moral law that is applicable in itself irrespective of the consequences of that action (Fig. 2.7).

---

[18] Immanuel Kant was a highly influential German philosopher who lived all his life in the German city of Koenigsberg in East Prussia. The German population were expelled from the city at the end of the Second World War (the city became part of Russia), and it is now called Kalingrad. Kant's most influential work "A Critique of Pure Reason" was a Copernican revolution in philosophy, and he rejected Hume's empiricism, and instead argued that some knowledge exists inherently in the mind independent of experience.

That is, Kant argues that *for people to behave ethically they must act from duty and obey the moral law*, and that the consequences of the act is not what makes it right or wrong, but rather the motives of the person performing the act. For Kant, something is "good in itself" when it is "intrinsically good", and it is "good without qualification" when its addition can never make the situation ethically worse. From this he concludes that *good will* is the only thing that is good without qualification.

Kant argues that the consequences of a person exercising their will may not be used to determine if a person has good will, as good consequences could arise by accident from an action intended to cause harm, whereas bad consequences could arise by accident from an action intended to cause good. Kant instead argues that a person has good will when they act out of respect for the moral law, when they act in a certain way when they have a duty to do so. Another words, the only thing that is truly good in itself is a good will, and a good will is only good when the person acts in accordance with duty out of respect for the moral law.

A moral law (according to Kant) is valid at all times and in all places, and is thus unconditionally applicable (*categorically* applicable) to everyone in all circumstances. Kant argued that there is one universal principle from which all moral laws or norms may be derived, and this principle is termed the *categorical imperative*, and is the foundation of all moral judgments. The categorical imperative is an absolute requirement that must be obeyed (i.e., the imperative is a prescribed action), and is an end in itself. That is, it is a law from which all duties and obligations arise.

Kant's formulation of the categorical imperative may be seen in the *universality principle* as follows: "*Act only on that maxim that you can at the same time will that it should become a universal law*". That is, the maxim (a practical principle that prescribes action) should be unconditionally good and should be able to serve as a general law for everyone without this leading to contradiction. Kant gives an example of someone making a false promise to repay money as this results in a contradiction on universalisation, and thus cannot become a universal law [6].

For Kant, the categorical imperative may also be expressed as a *law of nature*, where we should act in such a way that those maxims were to become universal laws of nature. Kant divides duties into two subsets: duties we have to ourselves versus duties we have to others, and we have a duty not to act on maxims that result in a logical contradiction.

Kant gives an alternate formulation of the categorical imperative in terms of the *reciprocity principle* as follows: "*Act only to treat humanity, whether in your own person or in that of any, in every case as an end and never as a means only*". That is, a moral action includes an end as well as a moral principle, and each human must have respect for the rationality of another, and must not misguide the rationality of another.

There are several criticisms of Kant's theory including that it is too rigid (i.e., there is no bending of the rules as Kant does not allow for exceptions in his theory), and that it ignores conflicts between moral laws. However, Kant remains highly influential in western philosophy.

2.8 Libertarianism

Libertarianism views individual freedom as the fundamental human value in society, and it regards any coercion or infringement of that liberty with deep suspicion. This means that libertarians wish to be left alone to be free to make choices about their lives and property, and they do not wish others (i.e., the state or government) to interfere with their right to liberty. This means that they are suspicions of distributive justice, where a democratic government seeks to redistribute wealth for the common good to ensure fairness in society, as well as essential services for its citizens (e.g., health, education, legal, police and defence). Libertarians tend to embrace individualism, self-reliance and a free market economy, and often view economic regulation as coercion and infringement of their rights, and curtailment of their freedoms in economic affairs.

Libertarian philosophy is an extreme free market or *laissez faire* philosophy that aims to get government out of things by reducing taxes, privatisation of government services, reducing or eliminating regulation and government supports, and increasing the role of the private sector in society. It rejects the view that everybody's needs should be represented in society, and instead argues for small government and against governments playing a role in reducing inequalities and redistributing wealth. It argues that taxes should be cut, as they infringe on an individual's freedom and liberty, which are fundamental goals for libertarians. Further, they argue that if taxes and government size are reduced then this will provide an incentive to entrepreneurs to create businesses, which will then lead to employment and benefits for workers, and to everyone being better off in society.

Libertarian politics became popular in the United States under President Ronald Reagan in the early 1980s. It has led to a massive transfer of wealth to the wealthiest of American society, and weaker regulation of American business (e.g., weaker regulation of the financial sector in Wall Street was a factor in the 2007–2008 global financial crisis). It has increased inequalities in the US with tax reductions for the wealthiest in society, and reductions in support for workers (especially in the social safety net programs). Further inequalities may be seen in the ratio of CEO pay (of large companies) to the average worker pay in the US, which was roughly 10–20 times average worker salary in the 1970s, whereas in recent decades it has increased to hundreds of times the average worker salary. This means that a fraction of 1% of Americans (0.01%) earn over 5% of American income, and there is a huge wealth gap between rich and poor in the United States.

2.9 Virtue Ethics

Utililitarianism and Kantianism are rule based ethical theories that are concerned with carrying out the right action in a given situation, whereas virtue ethics focus on the ethical nature of the person who is carrying out the action. That is, virtue ethics is concerned with building the right character traits in the person through education and good examples, as it is reasonable to expect that a person with good virtuous traits will perform good actions. Another words, the goal is to develop the person into a morally good and responsible human being who can lead a good life. Actions are good if they come from good traits of character, and the actions are guided by the desire of the individual to be virtuous.

The origins of virtue ethics are in ancient Greek philosophy especially in the work of Socrates, Plato and Aristotle, and the Stoics. Socrates was concerned with how one should lead a good life, and he saw wisdom as the path to virtue (*arete*) and the highest good (*eudaimonia* or happiness). Socrates emphasised that a good life is a good moral life, and he explored and questioned others in an attempt to give an account of *arete*. He emphasised that one should be conscious of how little one knows in order to learn more, and that one should think hard about the situation, and do the right thing is best for you. Further, if anyone does wrong it is because they have not thought enough about the situation, since, according to Socrates, no one does wrong willingly.

Aristotle argued that the final goal of human action is to strive for the highest good, and this involves achieving the state of being a good person and living a good and virtuous life. A moral virtue for Aristotle is a middle way between two extremes of evil, with for example, courage is balanced between cowardice and recklessness. Aristotle argues that people should seek a middle way, but that the middle course also depends on the circumstances in a given situation. This requires the intellectual virtue of practical wisdom to make the right moral judgment (which is the middle way), and so moral virtue and intellectual virtue go hand-in-hand.

The Stoics emphasised the importance of self-control as a way to overcome destructive emotions, and that rational thought and logic plays a key role in improving an individual's moral well being.

The key problem with virtue ethics is that it does not tell you how to act in a given situation, and it seems to be assumed that the virtuous person will always know the correct action in a given situation.

2.10 How Can an Individual Be Ethical in a Complex World?

Ethics provides guidance in making decisions to create just outcomes, and to have a positive impact on the world around us. Ethics arises in both our personal lives and in our work lives, and it provides us with a moral compass to guide our behaviour with respect to what is right or wrong in a particular situation that arises, and allows us to make an informed ethical decision.

An ethical person is responsible and accountable for his/her actions, and. does not cover up mistakes that have been made, and will instead be honest, responsible and trustworthy. There are several guiding principles in striving to be an ethical person such as treating people with dignity, being open and admitting mistakes, and acting in a way that one would like others to act towards you.

- Treating people with respect and dignity
- Behaving with honesty and integrity
- Avoiding hurting others
- Acting in a way that one would like others to act towards you
- Contributing to the betterment of others
- Admitting mistakes and moving on
- Awareness of legal and ethical responsibilities.

We mentioned the influential work of Fr. Bernard Lonergan in Chap. 1, and his four key steps on being human offers a useful path for leading an ethical life. These steps are *"Be attentive, be intelligent, be reasonable, and be responsible"*.

Ethical decision-making could also involve dealing with ethical dilemmas that may arise between two or more virtue-driven interests, and so one approach to deal with this may be to:

- Determine the facts
- Determine the ethical issues
- Determine options
- Evaluate options
- Choose option
- Implement solution.

2.11 How Can an Ethical Environment Be Created in an Organisation?

A business (or organisation) consists of a set of people each with individual values, and it is important that the personal values of the individuals are consistent with the code of ethics of the organisation. An organisation needs to create its mission, vision and values, and these should be communicated effectively within the organisation, and made an integral part of the work culture. The values include a set of ethical and moral standards that are appropriate for the organisation, and they need to be more than just window dressing to ensure that there is no unethical behaviour in the business practices of the workplace. Unethical behaviour should not be rewarded, and there needs to be structures in place to discipline employees who do adhere to the values of the organisation. The organisation needs to put a structure in place for employees to raise ethical concerns so that these may be dealt with in a professional manner. The organisation needs to:

- Define and communicate its values
- Make the values an integral part of the work culture
- Hold employees accountable for violations of its code of conduct

2.11 How Can an Ethical Environment Be Created in an Organisation?

- Reward good behaviour and discipline inappropriate behaviour
- Make organisation values part of the recruitment process.

An ethical organisation helps to promote a positive trustworthy view of the company, and helps to improve its reputation and its image. It leads to increased employee satisfaction and morale. The implementation of an ethical culture in an organisation places responsibilities on the employees to:

- Read organisation code of ethics document and sign it
- Attend training to become familiar with the organisation values and follow the values in the work place
- Report any conflicts of interest
- Follow all relevant laws
- Discuss any ethical dilemmas with management.

2.11.1 Ethical Decision Making

The process for solving moral and ethical problems is complex, as often such problems are poorly formulated or ill structured, and often there is a need to satisfy conflicting moral constraints. Their solution requires a clear formulation of the moral problem, analysis and identification of potential solutions, and the evaluation of the options to choose the most appropriate solution while balancing (conflicting) moral values. An ethical environment promotes a culture of trust, and the steps involved in ethical decision-making may include:

- Formulate the moral problem
- Analysis of facts and values with stakeholders
- Determine options on how to proceed
- Ethical analysis of available options
- Decision on the chosen option and rationale.

There is a difference between the legality and illegality of an act, and whether the act is ethical or unethical. The law is the arbiter on whether a particular act is legal or not, whereas the person's conscience or moral compass decides on whether a particular act is ethical or not. Table 2.1 presents some examples of the various combinations that may occur.

Table 2.1 Combinations of legal and ethical

Type	Example
Legal + ethical	This is the normal state of affairs where individuals and companies are obeying the law, and doing the right thing
Legal + unethical	An example is the failure of a manager to honour a commitment made to an employee, and this can lead to a breakdown in trust and respect in the manager/employee relationship
Illegal + ethical	Usually an illegal act is unethical, but an example could be a parent who is travelling just over the speed limit to bring a child to hospital for medical treatment
Illegal + unethical	An example would be a company breaking environment laws such as emissions, or pumping raw sewage into rivers or the sea

2.12 Review Questions

1. What is ethics?
2. Describe the main schools of ethics.
3. What is business ethics?
4. Give examples of unethical behaviour.
5. Explain the difference between utilitarian ethics and deontological ethics.
6. Explain the difference between Christian, Islamic and Buddhist ethics.
7. Explain how ethical decisions may be made.
8. How can an individual be ethical in a hostile work environment.

2.13 Summary

Ethics is a branch of philosophy that deals with moral questions such as what is right or wrong, and determining the right behaviour for an individual in a given situation. There are various schools of ethics such as the relativist position; cultural relativism; deontological ethics; and utilitarianism.

Business ethics is concerned with ethical principles and moral problems that arise in a business environment. They refer to the core principles and values of the organization. Professional ethics are a code of conduct that governs how members of a profession deal with each other and with third parties.

Ethics in Greek philosophy is concerned with rational thought and the search for appropriate principles to guide human conduct. It is characterised by its faith in reason to do the right thing. The great religious of the world provide a moral compass to their communities to behave ethically in the world, and we discussed the nature of ethics in Christianity, Islam, and Buddhism.

Utilitarianism was developed by Bentham in the nineteenth century, and its sole moral criterion for good or bad lies in the utility principle, which is the greatest happiness of the greatest number of a community. That is, the right action in a given situation is the action that produces the most benefit or happiness to society.

Deontological ethics argues that the ethical value of an action should be judged on whether the action is right or wrong in itself (based on a set of rules), and not on the actual outcome or consequences of the action. Virtue ethics is concerned with building the right character traits in the person through education and good examples, as a person with good virtuous traits will perform good actions. Actions are good if they come from good traits of character, and the actions are guided by the desire of the individual to be virtuous.

References

1. A. MacIntyre, *A Short History of Ethics* (Routledge, London, 1995)
2. T.J. Mawson, Belief in God. *An Introduction to the Philosophy of Religion* (Oxford University Press, Oxford, 2005)
3. B. Russell, W*hy I am not a Christian and Other Essays on Religion and Related Subjects*, 2nd edn. (Routledge, London, 2020) (Lecture delivered in 1927)
4. .J.R. Hinnells, *A Handbook of Living Religions* (Penguin, 1991)
5. S. Rinpoche, T*he Tibetan Book of Living and Dying* (Random House Ltd., London, 1998)
6. I. Kant, *Groundwork of the Metaphysics of Morals* (1785)

Professional Responsibility of Computer Professionals

3

Abstract

This chapter discusses the professional responsibilities of computer professionals, and we discuss the code of ethics of the Association for Computing Machinery, the Institute of Electrical and Electronic Engineers and the British Computer Society. We discuss the role of the whistle blower and its importance.

Keywords

Whistle blower • Code of ethics • IEEE code of ethics • ACM code of ethics • BCS code of ethics • Precautionary principle

3.1 Introduction

Professional responsibility refers to the responsibility of computer professionals to carry out their work professionally to the highest standards, and to use sound judgment in the exercise of their duties. Professionals are accountable to themselves and others for their actions, and they must be willing to accept professional responsibility when performance does not meet professional standards.

Parnas argues that software engineers have individual responsibilities as professionals. Another words, software engineers are responsible for designing and implementing high-quality and reliable software that is safe for the public to use. Engineers are accountable for their own decisions and actions, and should object to decisions that violate professional standards.

Professional engineers have a duty to their clients to ensure that they are solving the real problem of the client. Engineers need to be honest about current capabilities when asked to work on problems that have no appropriate technical solution, rather than accepting a contract for something that cannot be done.

Table 3.1 Professional responsibilities of software engineers

No.	Responsibility
1	Honesty and fairness in dealings with clients
2	Responsibility for actions
3	Continuous learning to ensure appropriate knowledge to serve the client effectively

The *licensing of a professional engineer* provides confidence that the engineer has the right education, experience to build safe and reliable products. Otherwise, the engineering profession gets a bad name because of poor work carried out by unqualified people. Professional engineers are required to follow rules of good practice and to object when rules are violated. The licensing of an engineer requires the engineer to complete an accepted engineering course, and understand the professional responsibility of being an engineer. The professional engineering body is responsible for enforcing standards and certification. The term *'engineer'* is a title that is awarded on merit, but *it also places responsibilities on its holder*.

Engineers have a professional responsibility and are required to behave ethically with their clients. The membership of the professional engineering body requires the member to adhere to the code of ethics of the profession. The professional responsibilities of software engineers include (Table 3.1).

3.2 What is a Code of Ethics?

A professional code of ethics expresses ideals of human behaviour, and it defines the core principles of the organisation. Several organisations such as the Association Computing Machinery (ACM), the Institute of Electrical and Electronic Engineers (IEEE), and the British Computer Society (BCS) have developed a code of conduct for their members. Violations of the code by members are taken seriously, and are subject to investigations and disciplinary procedures. A code of conduct for a professional body or corporation includes:

1. Guidelines for responsible behaviour of its members.
2. The guidelines may be prescriptive or a broad statement of values.
3. Violations of codes are investigated.
4. Members may be disciplined for violating the codes.
5. Professional codes are formulated by Engineering bodies
6. Corporate codes are formulated by companies.
7. Codes of conduct are an addendum to legal requirements.

There are various types of codes of ethics including (Table 3.2).

3.2 What is a Code of Ethics?

Table 3.2 Types of professional codes

Code	Responsibility
Aspirational codes	These are the values that the profession or company is committed to, and aspires to achieve
Advisory codes	These values help professionals to make moral judgments in different situations, based on the values of the profession or company
Disciplinary codes	These include disciplinary procedures to ensure that the behaviour of professionals adheres to the values specified in the code of ethics

A code of ethics places moral responsibility on computer professionals and software engineers to others and to society, and it includes ethical behaviour and responsibilities such as[1]:

1. Adhering to the values of the profession
2. Behaving with honesty and integrity
3. Fulfilling obligations to employer and to clients
4. Responsibility towards public and society.

The concept of corporate social responsibility (CSR) has become important in recent years, and this means that *companies have a responsibility to be good corporate citizens* in the societies in which they are operating. That is, they need to serve wider society in addition to their traditional commercial responsibilities to their shareholders. Another words, companies need to protect the environment, and must ensure the sustainability of their operations in addition to their traditional commercial responsibilities of protecting shareholder interest. That is, their additional responsibilities include environmental responsibility, ethical responsibility, philanthropic responsibility, and economic responsibility.

CSR can help to promote the corporation in the community where it operates, and to be seen as a socially responsible citizen in the community. It plays a role in ensuring that the corporation behaves ethically within society, and it has a positive impact on the environment.

Codes of conduct are values that members of a professional body or employees of a company are expected to adhere to, but may not be legally enforced as such. However, members of a particular profession or employees of a company that violate the codes may be subject to disciplinary procedures by the professional body or their employer. An effective code of ethics helps the corporation to achieve its corporate social responsibilities.

Unfortunately, codes of conduct may sometimes be *window dressing*, where the aspirations expressed in the code of ethics do not reflect the reality on the ground. The code may give the appearance that work is carried out in a certain

---

[1] These are core values of many mature software companies, and most companies operating today have a code of ethics that employees are expected to adhere to.

way (e.g., emissions below certain thresholds), and that the engineers are ethical in their day-to-day work. However, the reality on the ground may be quite different with unethical work practices taking place but being covered up. Further, codes of conduct have been criticised as being vague and contradictory, and this may create uncertainty for the employee or member of the professional as to what is the right action or behaviour is for a given situation.

Moral judgements and ethical decisions occur in various situations in a work environment, and so it would not be feasible for a code of ethics to cover all scenarios. In practice, a code of ethics expresses the moral principles of an organisation, and so an employee or software professional needs a *moral compass*, and to recognise situations where ethical decisions need to be made, and to apply their ethical judgment to a particular case.

There may be conflicts between the loyalty that a person has to their employer and their duty to do the right thing such as protecting the public. No employee desires to be placed in a situation where there is a conflict between what is morally right and their loyalty to their employer, and so it is important that organisations establish structures, where serious problems can be reported, discussed openly and dealt with appropriately. In rare situations, an employee may have no choice but to become a *whistleblower* to protect the public, where the organisation is intent on proceeding with a very risky approach that potentially endangers life or the environment. However, every effort should be made to avoid this situation, and this scenario is rare in practice (Fig. 3.1).

An employee may have a *conflict of interest* that could affect her professional judgment in a certain situation. For example, suppose that an employee has responsibility for selecting a new software package, and her husband runs one of the firms tendering for the work. Then an ethical employee would inform management of the conflict of interest, and remove herself from the selection process to remove any possibility of bias.

That is, a conflict of interest is an interest which if pursued interferes or conflicts with the obligation of the employee to his/her employer or client. The conflict of interest may corrupt or interfere with the employee's professional judgment, and it could potentially lead to inappropriate or immoral behaviour. It potentially destroys the trustworthiness of an individual, and so it is important to disclose a potential conflict of interest as soon as it arises.

Fig. 3.1 Whistle blower

Bribery and corruption are endemic to some countries, and as these are illegal activities (in most countries) the employee needs to report such activities when they arise. For example, an employee such as a purchasing manager is in a position of influence in an organisation, and could potentially be offered a *bribe* by another individual or company to influence his/her decision-making. Often, individuals or companies may be subtle in their attempt to gain influence on decision makers, with gifts or invitations to all-expenses paid events such as golf outings used to build up relationships with decision makers.

It is important to be cautious with respect to corporate entertainment, and many companies have policies that prohibit or restrict gifts to employees from external organisations or individuals. This helps to prevent employees being inappropriately influenced by others in their decision-making.

3.3 Role of a Whistle Blower

The whistleblower is a person who speaks out and informs the public on potentially unsafe or criminal acts in an organisation. However, speaking out should be the very last step in the process, as it could have serious consequences on the employee including career suicide or termination of employment. The steps involved in whistle blowing include first establishing the facts to determine the extent of the danger and its potential impact on the public, communicating the perceived danger and evidence for the danger within the organisation, and exhausting all internal procedures prior to acting. The whistle blower should only speak out when:

1. The organisation will do serious harm to the public.
2. The whistle blower has identified the threat, reported it to management, and concluded that management will not act.
3. The whistle blower has exhausted all internal procedures.
4. The whistle blower has convincing evidence that the threat is real.
5. The whistle blower believes that revealing the threat will prevent harm.

Table 3.3 describes the typical steps in whistle blowing.

Speaking out may be the ethical thing to do but often it comes at a serious cost to the employee, as he or she may be portrayed as being disloyal to the organisation. Further, as the organisation will wish to protect itself it may attempt to discredit the employee, and it may even terminate the employment of the employee. The organisation may portray the issue as a disgruntled employee whose employment was terminated due to performance issues with the employee's work.

Whistle blowing can also place a lot of emotional stress and strain on the employee, and even if the employee is not fired it may result in career termination in the organisation, with zero prospects of further promotion in the company. It is important that the employee protects himself by gathering all evidence on

Table 3.3 Steps in whistle blowing

No.	Responsibility
1	Establish the facts and double (or triple check) to ensure that you are factually correct with respect to the danger, and gather appropriate solid evidence that will convince any reasonable person of the danger
2	Report the matter and present the factual information to your immediate superior, and determine what action (if any) management will take
3	In the case of inaction escalate as appropriate within the organisation (organisations vary size/hierarchical structure and so escalation mechanism will differ) until all internal procedures are exhausted
4	In the absence of a reasonable resolution to the situation, or the organisation behaves in an unreasonable manner by failing to act or find an appropriate solution there may be no alternative but to speak out
5	The whistle blower reflects on the situation, weighs up the evidence and options, and decides that the only way to prevent harm is to speak out and reveal the danger to the public

the existence of danger, as this will be needed at a later stage. It may be prudent for the whistleblower to consider the consequences of speaking out and doing the right thing, both for themselves and others, to ensure that they fully understand the implications of the serious steps that they are taking, and that they will be able to manage the difficult circumstances in the aftermath of speaking out.

It may seem reasonable to suggest that an employee is fulfilling his moral duty if he informs management of the danger, as management are the decision makers with all the pertinent facts, and are thus best to make the final decision. However, such an approach can sometimes lead to loss of life, as with the Space Shuttle Challenger disaster back in 1986, which is discussed in Chap. 4.

Robert Boisjoly, an engineer at Morton Thiokel (a NASA subcontractor) was aware of the risks of erosion and failure when the 0-Rings of the Solid Rocket Booster (SRB) of the space shuttle are exposed to low temperatures. He argued that the space shuttle launch should not take place on the planned date due to the predicted temperatures, and advised management at Morton Thiokel of the situation. Boisjoly expected management to postpone the launch, but NASA placed pressure on Morton Thiokel to proceed with the launch, and the company stated that its data was inconclusive. Morton Thiokel gave its go ahead to proceed with the launch, which resulted in disaster and a serious loss of life of the crew of the space shuttle.

The IEEE code of ethics highlights the importance of speaking out in the case of danger, and it includes the code: *"disclose promptly factors that might endanger the public or the environment"*. The IEEE codes are discussed in the next section.

3.4 IEEE Code of Ethics

The Institute of Electrical and Electronic Engineers (IEEE) is the world's largest technical professional organisations with over 400,000 members in over 160 countries, and it is dedicated to advancing technology for the benefit of mankind. It publishes over 30% of the world's technical literature in electrical engineering, computer science and electronics as well as technical books and monographs. It is a leading developer of international standards in telecommunications and information technology, and individuals who have made outstanding contributions to engineering and technology may receive the prestigious IEEE Medal.

IEEE has developed a code of ethics for its members designed to ensure that they adhere to the highest ethical standards, and that its members treat others professionally and fairly to ensure that they are not discriminated against on the grounds of gender, race, and so on (Table 3.4).

The IEEE Code of Ethics requires its members to promptly disclose any factors that might endanger the public or society, which shows that it recognises the importance of whistle blowing and the need for members to speak out when there is danger to the public. The code mentions the importance of avoiding conflicts of interest and disclosing them when they occur, and stresses that unlawful activities such as bribery should be rejected. The code highlights the importance of carrying out roles only when one is qualified to do so, and to continue to improve one's technical competence. It emphasises that people should be treated fairly and with respect, without discrimination on gender, ethnicity, etc., and that harassment and injury to others should be avoided.

3.5 British Computer Society Code of Conduct

The British Computer Society (BCS) is a professional organisation for information technology and computer science that was founded by in 1957, and its first president was Sir Maurice Wilkes.[2] It has over 68,000 members in 150 countries, and it has played an important role in educating IT professionals. The BCS provides awards such as the Lovelace Medal[3] to individuals, who have made outstanding contributions to the computing field.

---

[2] Sir Maurice Wilkes developed the EDSAC computer at Cambridge University, which was one of the earliest stored-program computers. It was operational from May 1949.

[3] Ada Lovelace (the daughter of the poet, Lord Byron) was an English mathematician who collaborated with Charles Babbage (one of the grandfathers of the computing field) on applications of the Analytic Engine. Babbage designed the difference and analytic engines in the nineteenth century, and while the difference engine was a calculator for evaluating polynomials, the analytic engine was the design of the first mechanical computer.

Table 3.4 IEEE code of ethics

No.	Description
Highest ethical standards	
1	To hold paramount the safety, health, and welfare of the public, to strive to comply with ethical design and sustainable development practices, to protect the privacy of others, and to disclose promptly factors that might endanger the public or the environment
2	To improve the understanding by individuals and society of the capabilities and societal implications of conventional and emerging technologies, including intelligent systems
3	To avoid real or perceived conflicts of interest whenever possible, and to disclose them to affected parties when they do exist
4	To avoid unlawful conduct in professional activities, and to reject bribery in all its forms
5	To seek, accept, and offer honest criticism of technical work, to acknowledge and correct errors, to be honest and realistic in stating claims or estimates based on available data, and to credit properly the contributions of others
6	To maintain and improve our technical competence and to undertake technological tasks for others only if qualified by training or experience, or after full disclosure of pertinent limitations
Treating people fairly	
7	To treat all persons fairly and with respect, and to not engage in discrimination based on characteristics such as race, religion, gender, disability, age, national origin, sexual orientation, gender identity, or gender expression
8	To not engage in harassment of any kind, including sexual harassment or bullying behaviour
9	To avoid injuring others, their property, reputation, or employment by false or malicious actions, rumours or any other verbal or physical abuses
Following the code	
10	To support colleagues and co-workers in following this code of ethics, to strive to ensure the code is upheld, and to not retaliate against individuals reporting a violation

The BCS has developed a code of conduct that defines the standards expected of BCS members, and it applies to all grades of members during their professional work. Any known breaches of the BCS codes by a member are investigated by the BCS, and the appropriate disciplinary procedures followed. The main parts of the BCS code of conduct are listed in Table 3.5.

The BCS Code of Ethics requires its members to be conscious of the public health and environment. It states that one should only carry out those roles that one is qualified to do so, and one should continue to improve one's technical competence. It states the importance of avoiding conflicts of interest and that unlawful activities such as bribery should be rejected. It emphasises that members should seek to improve professional standards, and support other members in their professional development.

Table 3.5 BCS code of conduct

Area	Description
Public interest	Due regard to public health, privacy, security and environment
	Due regards to legitimate rights of third parties
	Conduct professional activities without discrimination
	Promote equal access to IT
Professional competence and integrity	Only do work within professional competence
	Do not claim competence that you do not possess
	Continuous development of knowledge/skills
	Understand/Knowledge / Comply with legislation
	Respect other viewpoints
	Avoid injuring others
	Reject bribery and unethical behaviour
Duty to relevant authority	Carry out professional responsibilities with due care and diligence
	Avoid conflicts of interest
	Accept professional responsibility for your work
	Do not disclose confidential information
	Accurate information on performance of products
Duty to the profession	Uphold reputation of profession and BCS
	Seek to improve professional standards
	Act with integrity
	Notify BCS if convicted of criminal offence
	Support other members in their professional development

3.6 ACM Code of Professional Conduct and Ethics

The Association of Computing Machinery (ACM) is the world's largest educational and scientific computing society, and it delivers resources that advance computing as a science. It has over 100,000 members around the world, and it includes a number of special interest groups (e.g., SIG AI is a special interest group on AI, and SIG SOFT is a special interest group on software engineering). The ACM has defined a code of ethics and professional conduct for its members, and the Code is summarised in Table 3.6.

The ACM Code of Ethics is comprehensive and requires its members to report any dangers that might cause damage or injury. The code mentions the importance of respecting intellectual property as well as privacy and confidentiality, and carrying out roles only when one is qualified to do so. Members should avoid conflicts of interest, and their work should be to the highest professional standards. Members should seek to improve their technical competence, and should treat people fairly and with respect. Finally, members should notify the ACM of any violations of the code.

Table 3.6 ACM code of conduct

No.	Area	Description
1. *General principles*		
1.1	Contribute to society and human well being	Computer professionals must strive to develop computer systems that will be used in socially responsible ways with minimal negative consequences
1.2	Avoid harm to others	Computer professionals must follow best practice to ensure that they develop high-quality systems that are safe for the public. The professional has a responsibility to report any signs of danger in the workplace that could result in serious damage or injury
1.3	Be honest and trustworthy	The computer professional will give an honest account of their qualifications and any conflicts of interest. The professional will make accurate statements on the system and the system design, and will exercise care in representing ACM
1.4	Be fair and act not to discriminate	Computer professionals are required to ensure that there is no discrimination in the use of computer resources, and that equality, tolerance and respect for others are respected
1.5	Respect property rights/intellectual property	The professional must not violate copyright or patent law, and only authorised copies of software should be made. The integrity of intellectual property must be protected, and credit for another person's ideas or work must not be taken
1.6	Respect the privacy of others	The professional must ensure that any personal information gathered for a specific purpose is not used for another purpose without the consent of the individuals. User data observed during normal system operation must be treated with the strictest confidentiality
1.7	Respect confidentiality	The professional will respect all confidentiality obligations to employers, clients and users
2. *Professional responsibility*		
2.1	Quality of processes/product	Computing professionals should strive to achieve the highest quality work throughout the process
2.2	Maintain high standards	It is essential to maintain high standards of technical knowledge and competence, and to upgrade skills on an ongoing basis

(continued)

Table 3.6 (continued)

No.	Area	Description
2.3	Respect rules	Computing professionals must adhere to rules including national and international laws and regulations
2.4	Professional review	Peer reviews play an important role in building quality into a work product, and computing professions should seek reviews of their work as well as participating in reviews
2.5	Comprehensive evaluations	Computing professionals are required to be thorough and comprehensive in their evaluation of computer systems including analysis and management of risk
2.6	Areas of competence	Computing professionals should only undertake work for which they have the required competence
2.7	Foster public awareness	Computing professionals should share technical knowledge with the public, and foster public awareness and understanding of computing
2.8	Authorised use of resources	Computing professionals should only access computer systems and software when they are authorised to do so
2.9	Secure systems	Computing professionals should develop robust and secure systems, as well as mitigation techniques and policies
3. Professional leadership		
3.1	Public good	The leader should ensure that the public good is the central concern during all professional computing work
3.2	Social responsibilities	Leaders should encourage computing professionals in meeting relevant social responsibilities
3.3	Quality of working life	Leaders should enhance the quality of working life of workers
3.4	Support principles of code	Leaders should pursue policies that are consistent with the code, and communicate them to the relevant stakeholders
3.5	Support growth of professionals	Leaders should ensure that opportunities are available to computing professionals to improve their knowledge and skill
3.6	Modifying/retiring systems	Leaders should exercise care when modifying or retiring systems
3.7	Special care	Leaders have a responsibility to be good stewards of systems that become part of the infrastructure of society

(continued)

Table 3.6 (continued)

No.	Area	Description
4. Compliance		
4.1	Uphold code	Computing professionals should adhere to the principles in the code and strive to improve them, and to express their concern to any individuals thought to be violating the code
4.2	Violations of code	ACM members who recognise a breach in the code should consider reporting the violation to the ACM

We discussed the professional responsibilities of some specific roles (e.g., the software engineer, the project manager and software tester) in our discussion of ethical software engineering in Chap. 1.

3.7 Precautionary Principle

The precautionary principle argues that if there is an identifiable risk of serious or irreversible harm, then it may be appropriate to place the burden of proof on the organisation proposing the potentially risky activity to show that it is safe, and for inaction until a proof of safety has been provided.

The main problem with the precautionary principle is that it potentially forbids too much, and opponents have argued that several innovations used today would not have been implemented if the precautionary principle had been adhered to. Further, its opponents argue that its demands for incontrovertible proof that there is no damage or harm is impractical, and that it is more sensible to demand that there are reasonable grounds for believing that there is no danger from the proposed activity or technology.

The precautionary principle may also be applied to unknown threats, where the principle permits preventive measures to be taken prior to fully knowing the seriousness of the threat. That is,

1. There is a threat
2. The threat is uncertain
3. Some kind of action is required
4. Action is taken.

3.8 Case Study on Workplace Ethics

The author has worked in both industry and academia, and in this section a small number of incidents that the author observed during his industrial experience are briefly discussed.

The first of these was an audio conference call that was made to a sister plant of the company in the United States, which the author attended with two to three other people. It was held in a senior manager's office, and when the conference call was about to end the senior manager gave the impression to the sister plant that he was about to leave the call, but instead he put the phone on silent and listened to the views expressed of the participants of the sister plant as to whether they expected his plant to deliver the project on time with the right quality. It allowed him to receive information that he would not otherwise have obtained, but it struck me at the time as being an underhand and inappropriate way of receiving that information.

The second incident was an arrangement made with a sister plant in the UK in relation to the weekly conference call between the plant in Ireland, the US and the UK. A written agreement was made between personnel in the plant in Ireland and the UK, where it was agreed that issues would be discussed first between the plant in Ireland and the UK prior to them being escalated to the US. The ink was hardly dry on the agreement before it was broken by the sister plant in the UK on the Friday conference call. And so, there was a sense of being let down by colleagues in another jurisdiction, and it led to a partial breakdown of trust that took time to heal.

3.9 Review Questions

1. Explain professional responsibility and accountability?
2. What is a code of ethics?
3. Describe the main features of the IEEE code of conduct.
4. Describe the main features of the BCS code of conduct.
5. Describe the main features of the ACM code of conduct.
6. What is the role of a whistle blower?
7. Give examples of conflicts of interest that could arise in the work place.
8. What is the precautionary principle?
9. What are the benefits to an organisation in having a code of ethics?

3.10 Summary

Professional responsibility refers to the responsibility of computer professionals to carry out their work professionally to the highest standards, and to use sound judgment in the exercise of their duties. Professionals are accountable to themselves and others for their actions, and they must be held accountable when performance does not meet professional standards.

Software engineers have professional responsibilities in that they are responsible for designing and implementing high-quality and reliable software that is safe for the public to use. They are also accountable for their own decisions and actions, and have a responsibility to object to decisions that violate professional standards.

Professional engineers are required to behave ethically with their clients. The professional responsibilities of software engineers include honesty and fairness in dealing with clients, responsibility for their actions, and continuous learning to ensure that they have appropriate knowledge to serve their clients effectively. The membership of the professional engineering body requires the member to adhere to the code of ethics of the profession.

A professional code of ethics expresses ideals of human behaviour, and it defines the core principles of the organisation. Several organisations such as the Association Computing Machinery, the Institute of Electrical and Electronic Engineers, and the British Computer Society have developed a code of conduct for their members. Violations of the code by members are taken seriously, and are subject to investigations and disciplinary procedures.

The precautionary principle argues that if there is an identifiable risk of serious or irreversible harm, then it may be appropriate to place the burden of proof on the organisation proposing the potentially risky activity to show that it is safe, and for inaction until a proof of safety has been provided.

Ethical Software Engineering

4

Abstract

This chapter discusses ethical software engineering and we discuss notable failures such as the space shuttle disaster and the defective Therac-25 radiotherapy machine. We discuss the extraordinary Volkswagen emissions scandal, where engineers designed a "defeat device" that would allow Volkswagen cars to pass emission tests in the United States.

Keywords

Safety and ethics • Therac-25 • Space shuttle disaster • Volkswagen scandal • Ethical project management • Bridge on the river Kwai • Ethical software testing • Ethical design and development

4.1 Introduction

Software engineering is a discipline that is concerned with the development of software, and it includes activities such as requirements gathering and definition, software design and development, and software testing to verify the correctness of the software. It is a team-based activity with several roles involved such as project managers, system analysts, developers and testers. Software engineering involves defining the right requirements, and then designing and implementing an appropriate solution that is fit for purpose, and verifying that the implementation satisfies the requirements.

Programmers are engineers and therefore need to learn the key engineering skills to enable them to build high quality products that are safe for the public to use. Software engineering involves the multi-person construction of multi-version programs. The IEEE 610.12 definition is:

Software engineering is the application of a systematic, disciplined, quantifiable approach to the development, operation, and maintenance of software; that is, the application of engineering to software, and the study of such approaches.

Software engineering includes:

1. Methodologies to design, develop, and test software to meet customers' needs.
2. Software is engineered. That is, the software products are properly designed, developed, and tested in accordance with sound engineering principles.
3. Quality and safety are properly addressed.
4. Mathematics may be employed to assist with the design and verification of software products. The level of mathematics employed will depend on the *safety critical* nature of the product. Systematic peer reviews and rigorous testing will often be sufficient to build quality into the software, with heavy *mathematical techniques reserved for safety and security critical software.*
5. Sound project management and quality management practices are employed.
6. Support and maintenance of the software is properly addressed.

Software engineering is not just programming. It requires the engineer to state precisely the requirements that the software product is to satisfy, and then to produce designs that will meet these requirements. The project needs to be planned and delivered on time and budget. The requirements must provide a precise description of the problem to be solved: i.e., *it should be evident from the requirements what is and what is not required.*

The requirements need to be rigorously reviewed to ensure that they are stated clearly and unambiguously, and reflect the customer's needs. The next step is then to create the design that will solve the problem, and it is essential to validate the correctness of the design. Next, the software code to implement the design is written, and peer reviews and software testing are employed to verify and validate the correctness of the software.

The verification and validation of the design is rigorously performed for safety critical systems, and it is sometimes appropriate to employ mathematical techniques for the verification. Mathematics plays a key role in classical engineering, and form part of the software engineer's toolbox for the safety critical field. Dijkstra and Hoare have argued that the way to develop correct software is to derive the program from its specifications using mathematics, and to employ *mathematical proof* to demonstrate its correctness with respect to the specification. Essential mathematics for software engineering are described in [1], but, it will often be sufficient to employ peer reviews and testing for verification and validation, as these methodologies provide a high degree of rigour.

The term *"engineer"* is a title that is awarded on merit in classical engineering. It is generally applied only to people who have attained the necessary education and competence to be called engineers, and who base their practice on sound engineering principles. The title places responsibilities on its holder to behave professionally and ethically. Often in computer science the term "*software engineer*" is employed loosely to refer to anyone who builds things, rather than to an individual with a core set of knowledge, experience, and competence.

Several computer scientists such as Parnas[1] have argued that computer scientists should be educated as engineers to enable them to apply appropriate scientific principles to their work. The use of mathematics is an integral part of the engineer's work in other engineering disciplines, and so the *software engineer* should be able to use mathematics to assist in the modelling or understanding of the behaviour or properties of the proposed software system.

Parnas has argued that software engineers have responsibilities as professional engineers. They are responsible for designing and implementing high-quality and reliable software that is safe to use. They are also accountable for their decisions and actions,[2] and have a responsibility to object to decisions that violate professional standards.

Engineers are required to behave professionally and ethically with their clients. The membership of the professional engineering body requires the member to adhere to the code of ethics of the profession. Engineers in other professions are licensed, and therefore Parnas argues that a similar licensing approach be adopted for professional software engineers to provide confidence that they are competent for the particular assignment.[3] Professional software engineers are required to follow best practice in software engineering and the defined software processes.[4]

Many companies invest heavily in training, as educated and knowledgeable employees are essential in the delivery of high-quality products and services.

---

[1] Parnas advocates a solid engineering approach with the extensive use of classical mathematical techniques in software development. He also introduced information hiding in the 1970s, which is now a part of object-oriented design.

[2] It is unlikely that an individual programmer would be subject to litigation in the case of a flaw in a program causing damage or loss of life. A comprehensive disclaimer of responsibility for problems rather than a guarantee of quality accompany most software products. Software engineering is a team-based activity involving many engineers in various parts of the project, and it would be potentially difficult for an outside party to prove that the cause of a particular problem is due to the professional negligence of a particular software engineer, as there are many others involved in the process such as reviewers of documentation and code and the various test groups. Companies are more likely to be subject to litigation, as a company is legally responsible for the actions of their employees in the workplace, and a company is a wealthier entity than one of its employees. The legal aspects of licensing software may protect software companies from litigation. However, greater legal protection for the customer can be built into the contract between the supplier and the customer for bespoke-software development.

[3] The British Computer Society (BCS) has introduced a qualification system for computer science professionals, which is used to show that professionals are properly qualified. The most important of these is the BCS Information Systems Examination Board (ISEB) which allows IT professionals to be qualified in service management, project management, software testing, and so on.

[4] Software companies that are following the CMMI or ISO 9001 standards will employ audits to verify that the processes and procedures have been followed. Auditors report their findings to management and the findings are addressed appropriately by the project team and affected individuals.

Employees need to receive appropriate training related to the roles that they perform, such as project management, software design and development, software testing, and service management. The fact that the employees are professionally qualified increases confidence in the ability of the company to deliver high-quality products and services. A company that pays little attention to the competence and continuous development of its staff will achieve poor results, and suffer a loss of reputation and market share.

There are several well-known lifecycles employed in software development such as the waterfall model, the spiral model, the Rational Unified Process, and the Agile methodology. The choice of a particular software development lifecycle for is determined from the needs of the specific project.

The waterfall model is employed for projects where the requirements can be identified early in the project lifecycle or are known in advance, and the phases include requirements, specification, design, and coding, unit tests, integration tests, system tests and acceptance testing [2]. Each phase has entry and exit criteria that must be satisfied before the next phase commences.

The spiral model was developed by Barry Boehm in the 1980s [3], and it is useful for projects where the requirements are not fully known at project initiation, or where the requirements evolve as a part of the development lifecycle. The development proceeds in a number of spirals, where each spiral implements new functionality, and typically involves determining objectives and analysing the risks, updates to the requirements, design, code, testing, and a user review of the particular iteration or spiral.

The *Rational Unified Process* [4] was developed at the Rational Corporation (now part of IBM) in the late 1990s. It uses the Unified Modelling Language (UML) as a tool for specification and design. UML is a visual modelling language for software systems that provides a means of specifying, constructing, and documenting the object-oriented system. James Rumbaugh, Grady Booch, and Ivar Jacobson developed it to facilitate the understanding of the architecture and complexity of a system.

There has been a massive growth of popularity among software developers in lightweight methodologies such as *Agile*. This is a methodology that is more responsive to customer needs than traditional methods such as the waterfall model. *The waterfall development model is similar to a wide and slow moving value stream, and halfway through the project 100% of the requirements are typically 50% done. However, for agile development 50% of requirements are typically 100% done halfway through the project.* Agile has a strong collaborative style of working [5, 6].

The CMMI is a framework to assist an organization in the implementation of best practice in software and systems engineering. It is used world-wide by thousands of organizations, and provides a solid engineering approach to the development of software. It supports the definition of high-quality processes for the various software engineering and management activities [7].

Software inspections play an important role in building quality into the software, and in reducing the cost of poor quality. There are several well-known inspection methodologies such as the Fagan Methodology [8] and Gilb's approach [9]. Software engineering is described in more detail in [10].

There is a need to make technical decisions quite frequently in software engineering, and often these decisions affect people's lives, and occasionally there are potential harmful impacts on others and on society. This means that the ethical impacts of technical decisions needs to be considered as part of the software engineering process, and so the ethical software engineer needs to examine both the technical and the ethical dimensions of decisions that affect wider society. An ethical software engineer should:

- Do no harm
- Do not take bribes
- Be fair to others.

A fundamental principle of ethics is based on the Hippocratic Oath "*Do no harm*", which may be seen to be breached where there are violations of ethics. For example, the Volkswagen emissions scandal discussed later in the chapter led to the deception of the general public and harm to society, the Volkswagen company, its shareholders and its employees. The actions of Volkswagen were unethical and illegal.

We discussed the professional responsibilities of software engineers in Chap. 3, as well as the code of ethics/conduct of several professional bodies such as IEEE, ACM and the BCS. The codes of ethics provide guidance on the interaction of technology and values, and software engineers need to be aware of their ethical responsibilities throughout the software development process, and to take action whenever ethical standards are in danger of being violated.

4.2 Safety and Ethics

The release of an unreliable software product may result in damage to property or injury (including loss of life) to a third party. Consequently, companies need to be confident that their software products are fit for purpose prior to their release. It is essential that software that is widely used is dependable, which means that the software is available whenever required, and that it operates safely and reliably without any adverse side effects.

Today, billions of devices and computers are connected to the Internet, and this has led to a growth in attacks on computers. It is essential that computer security is carefully considered, and that developers are aware of the threats facing a system, and of techniques to manage or eliminate them. The software developers need to be able to develop secure dependable systems that are able to deal with and recover from external attacks.

A safety critical system is a system whose failure could result in significant economic damage or loss of life. There are many examples of safety critical systems such as aircraft flight control systems, nuclear power stations and missile systems. It is essential to employ rigorous processes in the design and development of safety critical systems, and software testing alone is often insufficient in verifying the correctness of these systems.

The safety critical industry takes the view that any change to safety critical software creates a new program. The new program is therefore required to demonstrate that it is reliable and safe to the public, and so extensive testing needs to be performed. Additional techniques such as formal verification and model checking may be employed to provide an extra level of assurance in the correctness of the system.

Safety critical systems need to be reliable, dependable and available for use whenever required. The software must operate correctly and reliably without any adverse side effects. The consequence of failure (e.g., the failure of a weapons system) could be massive damage, leading to loss of life or endangering the lives of the public. We discuss two important case studies on disasters that occurred in the mid-1980s, and these are the Therac-25 disaster (see Sect. 4.2.1) and the Space Shuttle Challenger Disaster (see Sect. 4.2.2).

4.2.1 Therac-25 Disaster

The Therac-25 was a computer-controlled radiation therapy machine that was developed by the Atomic Energy of Canada (AECL) in the early 1980s. This linear accelerator treated cancer patients by exposing them to a beam of particles that would destroy malignant tissue (Fig. 4.1).

The machine consisted of hardware and software, and whereas the role of software on the earlier Therac-20 machine was limited, software played a more important role in the later Therac-25 machine. Its role was to perform many of the safety critical checks for the Therac-25, whereas this was performed by hardware on the earlier Therac-20 machine. The software on the Therac-25 radiation machine was responsible for:

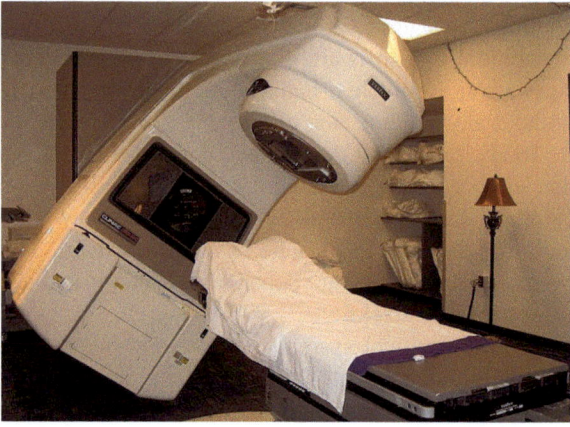

Fig. 4.1 A radiotherapy machine

4.2 Safety and Ethics

- Monitoring the status of the machine
- Accepting treatment input
- Setting up the machine for the treatment
- Turning on the treatment beam
- Turning off the treatment beam
- Detecting hardware malfunction.

There were six major accidents with the machine in the mid-1980s (1985–1987), where patients were given massive overdoses of radiation. The machine malfunctioned, and several patients received doses that were hundreds of time in excess of the appropriate dose, resulting in the death of three people and serious injuries to three others.

The machine continued in use for over 18 months after the first accident, with AECL believing that an accident was impossible with the machine, and it took no action with respect to the first reported accident. The second accident occurred a month later, and AECL sent an engineer on site to investigate the incident. He was unable to reproduce the problem, but AECL made some hardware and software changes and claimed that this solved the problem, as well as increasing the reliability of the machine a multiple of times.

AECL's response to the third accident was a denial of the problem, where they stated that the malfunction could not have been caused by the Therac-25 machine. They claimed that the 4th accident was as the result of a wiring problem. Finally, as a result of 5th accident, and FDA investigations into the operation of the Therac-25 machine, AECL finally launched a thorough investigation. The FDA ruled that the Therac-25 machines were defective, and advised AECL to prepare a corrective action plan, and to advise their customers of the problems with the machine.

The corrective action plan was prepared by AECL and presented to the FDA. It led to serious concerns in the FDA with respect to the software engineering practices employed in AECL, and the risks that these posed to the delivery of a high-quality product that would be safe to use and would not endanger the public. There was a lack of software engineering documentation for the software development, and the testing of the software was judged to be inadequate. The FDA directed AECL to do extensive testing on the system each time a small software change was made to ensure the safety of the software. The main reasons for the Therac-25 disaster include:

- Initial failure to believe end users
- Overconfidence of engineers in its correctness
- Poor software design and development practices
- Inadequate testing
- Poor resolution of software defects
- Software errors.

The Therac-25 disaster led to the deaths of three people and serious injury to three others. Software engineering practices were immature in the 1980s, but this is no excuse for what happened. It is basic common sense that a proper investigation should have been done after the first accident, and that all existing machines should have been considered to be unsafe until proved otherwise. That is, all Therac-25 machines should have been removed from operational use until the cause of the problem had been correctly identified, and appropriate solutions implemented to prevent a reoccurrence.

4.2.2 Space Shuttle Challenger Disaster

The Space Shuttle Challenger disaster is an important case study on engineering safety and workplace ethics. The disaster occurred in January 1986, when the space shuttle broke apart 73 s into its flight, and all of the seven members of the crew were killed. The Rogers Commission was formed to investigate the accident, and it found that the Challenger disaster was caused by a failure in the O-Rings sealing (a joint on the right solid rocket booster), where the record low temperatures had stiffened the rubber O-Rings and reduced their ability to seal the joint. The report also criticised the decision making process that led to the launch stating that it was deeply flawed, *with conflicts between engineering data and management judgments* (Fig. 4.2).

Robert Boisjoly, an engineer at Morton Thiokel launched strong objections to the launch, as he was aware of the risks of erosion and failure when the 0-Rings of the Solid Rocket Booster (SRB) are exposed to low temperatures. He argued that the shuttle launch should not take place on the planned date due to the predicted temperatures.

Both the NASA project team and the management team at Morton Thiokel had the opportunity to prevent the challenger disaster by postponing the launch. During the conference call on the evening prior to the launch the entire Morton Thiokel team recommended a postponement of the launch, as they recommended a minimum launch temperature of 52 °F. Temperatures were forecast to drop to 30 °F overnight which was likely to compromise the safety of the launch. They had expected NASA to rubber stamp the decision, but they were wrong, and NASA stated that the Morton Thiokel briefing was based on emotion rather than factual data. NASA requested Morton Thiokel to review their data again to determine if the data showed that it was unsafe to proceed, and the conference call was re-scheduled to later in the evening.

For a launch to take place all sub-contractors must sign-off on going ahead, and NASA seems to have encouraged (perhaps pressurised) Morton Thiokel to recommend the launch unless they could prove that it was unsafe to do so. The conference call had been delayed to allow Morton Thiokel management to consider all of the data, and the result of the Morton Thiokel management meeting (which excluded participation from Boisjoly) was to proceed with the launch. Morton Thiokel stated that its data was inconclusive at the conference call with NASA, and all sub-contractors agreed to proceed with the launch. Boisjoly later called the Morton Thiokel decision to go ahead to be unethical.

4.2 Safety and Ethics

Fig. 4.2 Space challenger disaster

Separatism is the idea that scientists and engineers provide the technical input and advice to management concerning a particular engineering situation, and management decide how best to proceed. That is, managers act as the decision maker taking all of the inputs into account to make a value judgment on the best way to proceed. This approach generally works fine in engineering, but problems arise when managers are trying to balance conflicting values such as achieving a strict delivery constraint and the safety of an operation, and where management believes (or encourages their subordinates to support their belief) that there is a small but manageable risk. It is essential to have openness and transparency in decision-making, where decisions are made based on the objective facts and data, and risks are kept to an absolute minimum and are manageable.

The *precautionary principle* requires that a particular course of action be demonstrated to be safe prior to being conducted. This was the normal *modus operandi* of NASA, but NASA changed the burden of proof the night before the launch to demand that Morton Thiokel prove to NASA management that it was unsafe to proceed with launch. However, once Morton Thiokel gave their approval and ignored the input of Robert Boisjoly, it could be argued that Boisjoly had a moral responsibility to be a whistle blower given the likelihood that safety would be compromised due to the forecasted low temperatures for the launch. Boisjoly may have taken the position that he had advised management of the dangers with the launch (following the principle of separatism), and that it was the responsibility of management to act by postponing the launch.

Either way, the flawed decision led to the deaths of all seven of the space shuttle crew, which included a schoolteacher, Christa McAuliffe, who was scheduled to be the first teacher in space. McAuliff was selected from the NASA Teacher in Space project, and she was scheduled to teach some lessons to students while in space. The Teacher in Space project was designed to inspire students, and to stimulate their interest in science and space exploration.

4.3 Ethical Project Management

Software projects have a history of being delivered late or over budget, and software project management is concerned with the effective management of software projects to ensure the successful delivery of a high-quality product, on time and on budget, to the customer. A project is *a temporary group activity designed to accomplish a specific goal such as the delivery of a product to a customer. It has a clearly defined beginning and end in time.*

Project management involves project planning and estimation; the management of resources; the management of issues and change requests that arise during the project; managing quality; managing risks; managing the budget; monitoring progress; taking appropriate action when progress deviates from expectations; communicating progress to the various stakeholders; and delivering a high-quality product to the customer.

Project managers are professionals and need to behave professionally and ethically at all times during the project. The *Project Management Institute* (PMI) has defined a code of ethics and professional behaviour for project management, which defines the expectations of the behaviour of project management professionals. The core values for the ethical conduct of project management professionals include:

- Professional responsibility
- Respect
- Fairness
- Honesty.

Project management professionals have a *responsibility* for the decisions that they make (or fail to make), and the actions that they take (or fail to take). They should accept only those assignments for which they have the required competence, and commitments made should be fulfilled. Errors or omissions should be corrected promptly, and any proprietary information provided should be protected. Further, any unethical or illegal conduct should be reported to management, and project management professions should be aware of the regulations and laws that govern their work.

4.4 Ethical Software Design and Development

Project managers have a duty to show *respect* to others including sensitivity of behaviour in working with others from different cultural backgrounds. This involves behaving professionally at all times, listening to others' point of view and seeking to understand them, and working through conflicts and disagreements with others.

Project managers have a duty to be *fair* in decision making with decisions made objectively and impartially, and they should refrain from participating in decision-making where there is a potential conflict of interest. Further, favouritism and discrimination are not allowed.

Finally, it is the duty of project managers to act in a truthful and *honest* manner in their communication and conduct, and not to engage in or condone behaviour that attempts to deceive others (e.g., making misleading or false statements).

The project manager is accountable for the success of the project, and larger projects have more opportunities for ethics being compromised than smaller projects. Project managers endeavour to balance budget, schedule, effort and quality, which may potentially lead to ethical dilemmas when the project manager is tempted to cut corners to enable the project to be delivered on time and on budget. This could potentially result in quality being compromised, health and safety being compromised, privacy being compromised, and so on.

The selection of a subcontractor could pose a conflict of interest to the project manager, where the project manager knows one of the candidate subcontractors from a previous working relationship or family relationships. It is therefore important that in such a situation that the project manager excludes herself from the supplier selection to ensure that there is no conflict of interest.

Project management involves ethical decision-making, and good project governance is a good enabler of ethical project management. It enables the key project stakeholders to be kept informed of the key project status and the key decisions being made regularly during the project.

4.4 Ethical Software Design and Development

Ethical software design and development is concerned with ethical issues that may arise during technology development, such as questions as to how the technology will be used, and whether it could lead to harm to individuals and society. The design of a technology determines how it will be used, and this means that there needs to be an ethical dimension to the design process, where ethical values are considered as well as the desired functionality.

David Lean directed the movie *"The Bridge on the River Kwai"* in 1957,[5] and the film was based on the historical construction of the Thailand-Burma railway during the Japanese occupation of Burma in the Second World War. British prisoners of war are ordered to construct the bridge, and initially the British and their leader,

---

[5] The film was actually made on the Kelani river near Kitulgala in Ceylon (Sri Lanka).

Fig. 4.3 Bridge over the River Kwaii in Kanchanaburi

Colonel Nicholson, resisted participation in its construction. However, Colonel Nicholson later becomes obsessed with designing and building a proper bridge that will last well beyond the war, and that will be a tribute to the skill and ingenuity of British engineers (Fig. 4.3).

They build a solid bridge over the river and on the day that it is due to be opened with the first train due to pass over Nicholson finally realises the gravity of what he has done (i.e., collaborating with the enemy and contributing to their plans for further aggression). He blows up the bridge sending the train into the river. That is, *the purpose of the technology* (i.e., the completed bridge) needed to be considered, as a completed bridge would cause harm to others in that it would have facilitated an expansion of Japanese aggression to other countries. Further, it was unethical for Nicholson to collaborate with his enemy who wished to harm him and his country, and so his collaboration conflicted with his duties to his country and to the British army.

Software design is the process where certain functions are translated into a blueprint for a system that can fulfil these functions. There are often several design choices for a particular technology, and different designs may vary in the extent to which they deal with particular ethical values. The goal is to choose the design that best meets the most important ethical values and technology considerations, and this means that responsible choices must be made in the selection of the most appropriate design. Software design is a systematic process that uses technical and scientific knowledge, and there may need for trade offs with conflicting ethical values in the different designs. It involves activities such as:

4.4 Ethical Software Design and Development

- Problem analysis
- Requirements analysis and definition (may include prototyping)
- Architectural Design (may include design options and decision)
- Low Level Design
- Implementation
- Testing
- Maintenance.

Value centred design (VCD) is an approach to design that involves taking human values into account during the design process, and solving value conflicts through engineering design and technological innovation. It involves investigating and determining the values that are relevant to the project, and understanding conflicts to make tradeoffs. There is a need to analyse designs to determine the extent to which they meet particular values, and to develop innovative designs to meet particularly relevant values. Valued centred design involves:

- Reasoning/clarifying values underlying conflicting design requirements
- Social cost benefit analysis (including monetary costs for safety)
- Evaluation criteria (including value criteria, weightings may be employed)
- Thresholds for what is acceptable for each criterion
- Evaluation of options
- Selected option.

There may be conflicts between ethical values when choosing between two or more design options, and where the different designs score well on different criteria. This is where designers are unable to do justice to all ethical values simultaneously, and often the resolution of these moral dilemmas require a trade off or balancing between competing values. A trade off decision is where a choice needs to be made between at least two options, in which at least two moral values are relevant as choice criteria, and so finding the right balance in the trade off decisions may be a challenge (Fig. 4.4).

Software designers have a responsibility to create ethical designs that satisfy the requirements, and to ensure that their designs are robust and protect the safety of the public. Ethics is an important design concern that should be considered, and. this will determine how well the product fits within the ethical boundaries. There may be several ethical values that may be relevant, including safety, accessibility, usability, sustainability, privacy, security, honesty, fairness and loyalty. The evaluation of each design option should rate the extent to which the relevant ethical values are addressed by that option as well as the technical criteria.

Data management is an important part of ethical software engineering, where personal data ownership as well as data rights, access rights, privacy and security rights need to be considered and protected. Software designers need to follow the legal requirements as well as best practice in privacy and security in collecting,

Fig. 4.4 Balancing an ethical life against a feather in Egyptian Religion

processing and protecting data. An ethical system needs to be accessible, and its design should consider its accessibility for different categories of users, such as those with visual or hearing impairments, or those with different levels of language ability or education.

The ethical design of a software system should give an open and accurate account of the system, and should satisfy all relevant legal and regulatory requirements. We discuss the Volkswagen *Dieselgate* emissions scandal in the next section, where the unethical conduct of management of the company involved tasking software designers to develop software to create a "defeat device" that would enable Volkswagen vehicles to cheat the federal emissions tests in the United States.

Ethical software designers need to be conscious of the algorithms that they create to ensure that they are unbiased, and do not discriminate against minority groups in society. This is especially important in algorithms based on pattern matching that are employed in the AI field, where *biased algorithms* may lead to discrimination against individuals and groups. There have been several controversial algorithms such as the infamous Amazon hiring algorithm that discriminated against females, and a predictive policing algorithm that led to racial profiling and discrimination against racial groups. These issues are discussed in more detail in Chap. 6.

Software designers should consider the ultimate purpose of the project including its benefits to society as well as harm from the technology. We discussed the purpose of the bridge over the river Kwai, and argued that its design and implementation had the potential to harm Britain and the Allies in their efforts to defeat

4.4 Ethical Software Design and Development

the Axis powers. Social media and various other apps are deliberately designed to be *addictive* to their users, where the software captures the attention of the human at a primal level, and the company reaps financial gain from the addiction of the users. Humans have become addicted to their smartphones, and check their phone hundreds of time a day, and their addiction has been caused by addictive software design. This poses questions on the ethics of this addictive design, and whether the consequences of design and the end product should be considered in ethical decision-making.

Software companies also need to be environmentally responsible in the partnerships that they develop with other companies, especially with respect to the ultimate purpose of the partnership. For example, the Big Oil industry is active in climate pollution, and several cloud-computing companies have worked with oil companies and used AI and machine learning technologies to create visual maps to enable them to determine where best to drill to extract more polluting fossil fuels. Another words, some software companies turn a blind eye as to how its software products are used, and it is essential that a company committed to environmental sustainability behaves responsibly.

The system needs to be designed for security, as it is difficult to add security after the system has been implemented. Security engineering is concerned with the development of systems that can prevent malicious attacks, and recover from them. Software developers need to be aware of the threats facing a system, and develop solutions to manage them. Security loopholes may be introduced in the development of the system, and so care needs to be taken to prevent these as well as preventing hackers from exploiting security vulnerabilities.

There is a need to conduct a risk assessment of the security threats facing a system early in the software development process, and this will lead to several security requirements for the system. That is, the requirements of the system should specify security and privacy requirements, and the software design and development must implement them to ensure that security and privacy are not breached. Security testing (including penetration testing) is carried out to identify any flaws in the security mechanisms of the computer system, and to verify that the security requirements such as confidentiality, availability, integrity, etc., are satisfied. However, the successful completion of security testing does not guarantee that there are no security vulnerabilities in the system. Hackers will still attempt to steal confidential data and to disrupt the services being offered by a system.

4.4.1 Volkswagen Emissions Scandal

The Volkswagen *Dieselgate* scandal arose as a result of the German company deliberately programming its turbocharged direct injection (TDI) diesel engines to activate their emissions controls only during laboratory emissions tests. This meant that the vehicles NO_x emissions passed the US federal regulatory requirements during laboratory tests, whereas the actual emissions were over 40 times higher in real-world driving (Fig. 4.5).

Fig. 4.5 Volkswagen beetle type 82E

Volkswagen deployed this software in over 11 million vehicles world wide including roughly half a million vehicles in the United States from 2009 to 2015. It became evident in 2014 that there were discrepancies in emissions between European and US models, and regulators in several countries launched an investigation into Volkswagen. Several senior executives resigned or were suspended, and Volkswagen spent billions in recalling the affected vehicles and rectifying the issues with the emissions.

Volkswagen pleaded guilty to criminal charges in 2017, and they admitted to developing a "*defeat device*" to enable diesel models to pass US emission tests and deliberately concealing its use. Volkswagen was fined $2.8 billion for rigging the vehicles to cheat on the emission tests. The scandal had cost Volkswagen $33 billion in fines, penalties, financial settlements and buyback costs by mid-2020. Martin Winterkorn resigned his position of the CEO of Volkswagen in 2015, and he was charged with fraud and conspiracy in the United States in 2018.

The scandal highlighted how software controlled machinery is prone to cheating, and it has opened up a debate on whether there is a need for a mechanism to independently verify software that is employed to satisfy legal or regulatory requirements. That is, should all such software code be published for scrutiny by independent regulators and/or independently certified?

The Volkswagen scandal is deeply concerning as it demonstrates the failure of corporate/business ethics to act as a barrier to the pursuit of business self-interest. Volkswagen is a prestigious German company and it is extraordinary that the professionalism that Germany is renowned for could be tarnished in this way. Unfortunately, sometimes the code of ethics of an organisation are just window dressing for the public, rather than being embraced and engrained in the day to day work practices of corporate life.

Why did engineers fail to consider their ethical responsibilities? Why did they fail to question the implementation of this device? Why were there no whistle blowers to speak out against these unethical practices? Was there a lack of moral courage among the engineers? Were there appropriate structures in place for whistleblowers to discuss ethical concerns? Was there a total failure of corporate governance? Volkswagen's actions were illegal and deeply unethical, and its good name has been deeply tarnished. Herbert Diess became CEO in 2018, and helped to restore Volkswagen's tarnished reputation as well as overseeing the transition to electric vehicles.

A corporate environment is generally focused on the business and product implementation rather than on critical reflection on the wider implications of the technology. Engineers are often busy with their lives outside the office while trying to build a career within the office, and speaking out may not be viewed as a career-advancing move. Further, a hierarchical work environment does not actively encourage speaking out on issues outside of product development, with corporate enterprises often command driven operations, with power assigned within the hierarchy, and subordinates may fear the consequences of speaking out.

Engineers are often focused on getting the software to perform correctly to meet its specification, and so often may not consider the wide societal impacts of the technology. However, it is in the interest of both corporations and their employees to consider the bigger picture, and to actively consider ethical issues in the design process. Otherwise, they could well pay the price for their inaction later with significant damage to the reputation of the corporation and financial loss.

4.5 Ethical Software Testing

Software testers are professionals and need to behave ethically at all times during testing. The ISTQB Code of Ethics for test professionals is based on the IEEE and ACM code of ethics and it states that:

- Certified software testers shall act consistently in the public interest.
- They act in the best interests of their client and employer.
- They ensure that their deliverables meet the highest professional standards.
- They maintain independence and integrity in professional judgments.
- They shall promote an ethical approach to the management of software testing.
- They shall advance the integrity and reputation of the profession.
- They shall be supportive of colleagues and cooperate with software developers.
- They shall participate in lifelong learning and promote ethics in their profession.

Comprehensive testing reduces the risk of serious quality problems with the software, but it is impossible to test everything due to time constraints. This means that the testers need to focus their testing on the areas of greatest risk with the software, and on the parts of the system that the users are most likely to be using.

It is essential that the testers have the appropriate expertise, that the right test environment is set up, that they have prepared test plans and test specification to test the software, and that they have all the required tools in place.

Ethical issues may arise during testing if the project is behind schedule, and when there is pressure applied to the test team to stay with the original project delivery schedule. It may be that the available time for testing is insufficient to verify the correctness of the software, or the limited time could lead to testers missing serious defects. This could lead to the quality of the released software being compromised, and the test manager needs to resist any pressure that poses risks to quality, and needs to raise concerns at senior level where appropriate.

It is essential that the customer be informed of all quality problems with the software to ensure that they can manage any associated risks. The final test report should summarise the testing that has been done, the results of the testing, the open problems, the problem arrival rate, and known risks with the software. The final test report generally includes a recommendation from the test manager to release the software, and such a recommendation should be based on the key facts with a clear statement that all risks can be managed.

There may be conflicts when the project manager wishes to release the software on schedule, and where the test manager has concerns or believes that it is unsafe to do so based on the key testing status and risks. It is essential in such situations that the decision made is based on the facts and risks, and objective data should support the decision.

4.6 Review Questions

1. What is ethical software engineering?
2. Explain how the Therac-25 disaster occurred.
3. Explain how the challenger disaster occurred.
4. What is ethical software design?
5. What is value centred design?
6. What is ethical software testing?
7. What is ethical project management?
8. Explain the concept of separatism?
9. What are the ethical considerations in the development of safety critical systems?
10. Explain the ethical considerations in the Volkswagen dieselgate scandel.
11. Explain the ethical considerations in David Lean's film "Bridge over the river Kwai".

4.7 Summary

Technical decisions regularly need to be made regularly in software engineering, and as these decisions may have wider impacts it is important to consider their ethical dimension as part of the software engineering process. That is, the ethical software engineer needs to examine both the technical and the ethical dimensions of decisions that affect wider society.

A fundamental principle of ethics is based on the Hippocratic Oath "Do no harm", which may be seen to be breached where there are violations of ethics. For example, the Volkswagen emissions scandal was unethical and illegal and led to the deception of the general public and harm to wider society.

We discussed the defective Therac-25 computer-controlled radiation therapy machine that was developed by the AECL in the early 1980s. There were several major accidents with the machine in the mid-1980s, where patients were given massive overdoses of radiation.

We discussed the space shuttle challenger disaster in the mid-1980s, and the disaster was caused by a failure in the O-Rings sealing on the right solid rocket booster. The decision making process that led to the launch was deeply flawed, with conflicts between engineering data and management judgments.

Software designers need to consider the ultimate purpose of the project including its benefits to society as well as potential harm of the technology. We discussed the David Lean film concerned with building a bridge over the river Kwai, argued that it was unethical as it involved collaboration with the enemy and contributing to their plans for further aggression.

Value centred design involves taking human values into account during the design process, and solving value conflicts through engineering design and technological innovation. It involves investigating and determining the values that are relevant to the project, and understanding conflicts to make tradeoffs. There is a need to analyse designs to determine the extent to which they meet particular values, and to develop innovative designs to meet particularly relevant moral values.

References

1. G.O' Regan, *Mathematical Foundations of Software Engineering* (Springer, Berlin, 2023)
2. W. Royce, The software lifecycle model (waterfall model), in *Proceedings of WESTCON*, Aug 1970
3. B. Boehm, A spiral model for software development and enhancement. *Computer*, May 1988
4. J. Rumbaugh et al., *The Unified Software Development Process* (Addison Wesley, New York, 1999)
5. K. Beck, *Extreme Programming Explained. Embrace Change* (Addison Wesley, New York, 2000)
6. K. Beck et al., *Manifesto for Agile Software Development*. Agile Alliance. http://agilemanifesto.org/ (2001)

7. M.B. Chrissis, M. Conrad, S. Shrum, CMMI for development. Guidelines for process integration and product improvement, 3rd edn. SEI Series in Software Engineering (Addison Wesley, New York, 2011)
8. M. Fagan, Design and code inspections to reduce errors in software development. IBM Syst. J. **15**(3) (1976)
9. T. Gilb, D. Graham, *Software Inspections* (Addison Wesley, New York, 1994)
10. G. O' Regan, *Concise Guide to Software Engineering*, 2nd edn. (Springer, Berlin, 2022)

Ethical and Legal Aspects of Data Science

5

Abstract

This chapter discusses ethical and legal aspects of data science. There has been a phenomenal growth in the use of digital data with vast amounts of data collected, processed and used, and so the ethics of data science has become important. It aims to investigate what is fair and ethical in data science, and what should or should not be done with data. We discuss ethical problems of privacy that arise in social media, the Internet of Things, AI and facial recognition, and the legal aspects of privacy.

Keywords

Data science • Data scientist • Informed consent • GDPR • DSA • Privacy • Security • AI • Internet of things • Social media

5.1 Introduction

Companies collect lots of personal data about individuals, and so the question is how should a company respond to a request for personal information on particular users? Does it have a policy to deal with that situation? What happens to the personal data that a bankrupt company has gathered? Is the personal data part of the assets of the bankrupt company and sold on with the remainder of the company? How does this affect privacy information agreements and compliance to them or does the agreement cease on termination of business activities?

The consequence of an error in data collection or processing could result in harm to an individual, and so the data collection and processing needs to be accurate. Decisions are often made on the basis of public and private data, and individuals may be unaware as to what data was collected about them, whether the data is accurate, and whether it is possible to correct errors in the data.

Further, the conclusions from the analysis may be invalid due to errors in incorrect or biased algorithms, and so a reasonable question is how to keep algorithmically driven systems from harming people? Data scientists have a responsibility to ensure that the algorithm is fit for purpose and uses the right training data, and as far as practical to detect and eliminate unintentional discrimination in algorithms against individuals or groups.

That is, problems may arise when the algorithm uses criteria tuned to fit the majority, as this may be unfair to minorities. Another words, the results are correct, but presented in an over simplistic manner. This could involve presenting the correct aggregate outcome but ignoring the differences within the population, and so leading to the suppression of diversity, and discriminating against the minority group. Another example is where the data may be correct but presented in a misleading way (e.g., the scales of the axis may be used to present the results visually in an exaggerated way).

This chapter is concerned with the ethics of data Science, and we discuss data science in more detail in the next section.

5.2 Data Science

Information is power in the digital age, and the collection, processing and use of information needs to be regulated. Data science involves the extraction of knowledge from data sets that consist of structured and unstructured data, and data scientists have a responsibility to ensure that this knowledge is used wisely and not abused. Data science may be regarded as a branch of statistics as it uses many concepts from the field, and in order to prevent errors occurring during data analysis it is essential that both the data and models are valid.

Personal data is collected about individuals from their use of computer resources such as their use of email, their Google searches, as well as their Internet and Social media use to build up revealing profiles of the user that may be targeted to advertisers. Data monetisation (or surveillance capitalism) is concerned with the commercial exploitation of data gathered from the surveillance of users (e.g., their search history, browsing history and so on), and it involves the widespread collection and commodification of personal data collected by corporations. Modern technology has allowed governments to conduct mass surveillance on its citizens, with face recognition software allowing citizens to be recognised at demonstrations or other mass assemblies.

Smartphones provide location data that allows the location of the user to be tracked, and so it is important that such technologies are regulated and not abused by the state. Privacy has become more important in the information age, and it is the way in which we separate ourselves from other people, and is the right to be left alone. The European GDPR law has become an important protector of privacy

5.2 Data Science

Table 5.1 Sources of information

Source	Description
Data collected by merchants and service providers	This includes personal data entered for the purchase of products and services such as name, address, date of birth, products and services purchased, etc.
Activity tracking	This involves monitoring the user's activity on the site, recording their searches, and the products that are browsed and purchased It may involve recording the user's interests, their activities, and their interactions / communications
Search Profile	The history of a person's searches over a period of time on a search engine such as Google reveals information about the individual and their interests
Sensors from devices	These include personal devices that are part of the Internet of Things that may record information such as health data or what the individual is eating Security cameras may be conducting public or private surveillance. GPS technology on smart phones may monitor the user's location

and personal data, and European and other countries have adapted it. The main sources of personal data that are collected include (Table 5.1).

These sources of information can collect vast amounts of data, and the collected data may potentially result in harm to an individual. The collected data is commercially valuable, especially when data about individuals are linked from several sources. *Data brokers* are companies that aggregate and link information from multiple sources to create more complete and valuable information products (i.e., profiles of individuals) that may then be sold on to interested parties. Meta data (i.e., data about the data such as the time of a phone call or who the call is made to) also provides useful information that may be collected and shared.

For example, suppose that the probability of an individual buying a pair of hiking books is very low (say 1 in 5000 probability). Next, that individual starts scanning a website (say Amazon) for boots then that individual is now viewed as being more likely to buy a pair of hiking boots (say a 1 in 100 probability). This large increase in probability will mean that the individual is now of interest to advertisers and sellers, and various targeted (popup) advertisements will appear advertising various hiking boots to the individual. This may become quite tedious and annoying to the individual, who may have been just browsing, and is now subject to an invasion of advertisements. However, many apps are free and often their source of revenue is from advertisements, and so they gather data about the user that is then sold on to advertisers.

5.3 Ethics of Data Science

There has been a phenomenal growth in the use of digital data in information technology, with vast amounts of data collected, processed and used, and so the ethics of data science has become important. There are social consequences to the use of data, and the ethics of data science aims to investigate what is fair and ethical in data science, and what should or should not be done with data.

Ethical behaviour is part of daily life, and it refers to how we use our internal moral compass to form judgments and make decisions in various ethical situations. For example, if I take a picture of another individual does the picture belong to me (as owner of the camera and the collector of the data)? Or does it belong to the individual who is the subject of the image? Most reasonable people would say that the image is my property, and if so, the question is what responsibilities or obligations do I have (if any) to the other individual?

That is, although I may technically be the owner of the image, the fact that it contains the personal data (or image) of another should indicate that I have an ethical responsibility or obligation to ensure that the image (or personal data) is not misused in any way to cause harm to that individual. Further, if I misuse the image in some way then I may be open to a lawsuit from the individual.

Ethical rules are shared values that are followed voluntarily to make the world a better place, whereas legal rules are used to enforce social values. Often, the benefits of following the rules outweigh the costs of following them. For example, following the defined rules of the road lead to safe and predictable travel, whereas the cost of obeying the rules is that an individual must conform to driving practices of the particular country, including driving under the speed limit and driving on the correct side of the road as well as following the rules of the road.

A fundamental principle of ethics in data science refers to *informed consent*, and this has its origins in the ethics of medical experiments on individuals. The concept of informed consent in medical ethics is where the individual is informed about an experiment, and gives their *consent voluntarily*. The individual has the right to withdraw consent at any time during the experiment. Such experiments are generally conducted to benefit society, and often there is a board that approves the study and oversees it to ensure that all participants have given their informed consent, and attempts to balance the benefits to society with any potential harm to individuals. Once individuals have given their informed consent data may be collected about them.

The principle of informed consent is part of information technology, in the sense that individuals accept the terms and conditions of a website before they may use the software, and these terms state that data will be collected, processed and shared. However, it is important to note that users do not generally give informed consent in the sense of medical experiments, as the details of the data collection and processing is hidden in the small print of the terms and conditions, and this is generally a long and largely unreadable document. Further, the consent is not given voluntarily, in the sense that if a user wishes to use the software, then he or she has no choice but to click acceptance of the terms and conditions of use for the site. Otherwise, they are unable to access the site, and so for many website and software applications (apps) *consent is essentially coerced rather than freely given.*

There was some early research done on user behaviour by Facebook in 2012, where they conducted an experiment to determine if they could influence the mood of users by posing happy or sad stories to their news feed. The experiment was done without the consent of the users, and while the study indicated that happy or sad stories did influence the user's mood and postings, it led to controversy and major dissatisfaction with Facebook when users became aware that they were the *subject of a psychological experiment without their consent.*

The dating site OKCupid (a US and internationally based dating and friendship site) uses an algorithm to find compatibility matches for its users based on their profiles, and two people are assigned a match rating based on the extent to which the algorithm judges them to be compatible. OKCupid conducted psychological experiments on its users without their knowledge, with the first experiment being a *"love is blind"* day in 2014, where all images were removed from the site, and so compatibilities were determined without the use of images.

Another experiment was more controversial and unethical, as the site lied to the users on their match ratings (e.g., two people with a compatibility rating of 90% were given a rating of 30%, and vice versa). The site was trying to determine the extent that two people would get along irrespective of the rating that they were given, and it showed that two people talked more when falsely told that the algorithm matched them, and vice versa. The controversy arose once users became aware of the deception by the company, and it provides a case study on the *socially unacceptable manipulation of user data* by an Internet company.

Data collection is not a new phenomenon as devices such as cameras and telephones have been around for many years. People have reasonable expectations on privacy, and do not expect their phone calls to be monitored and eavesdropped by others, or they do not expect to be recorded in a changing room or in their home. Individuals will wish to avoid harm that could occur due to data about them being collected, processed, and shared. The question is whether reasonable rules can be defined and agreed, and whether tradeoffs may be made to balance the conflicting rights of the business and individual, and to protect the individual as far as is possible. Some questions on data collection and ownership are considered in Table 5.2.

5.3.1 Problems in Data Science

Data science is a multi-disciplinary field that extracts knowledge from data sets that consist of structured and unstructured data, and large data sets (*big data*[1]) may be analysed to extract useful information. The field has great power to harm

---

[1] Big data involves combining data from lots of sources such as bar codes, cctv, shopping data, drivers license, and so on.

Table 5.2 Questions on data collection

Question	Answers
Who owns the data?	A user's personal information may legally belong to the data collector, but the data subject may have some control as the data is about him/her • The author of the biography of an individual owns the copyright not the individual • The photographer of a (legally taken) photo owns the image not the subject • Recording of audio/video is similar • May be a need to acknowledge copyright (if applicable) • May be limits in rights as to how data is collected and used (e.g. privacy of phone calls) • The data subject may have some control of the data collected
What are the expected responsibilities of the data collector	The collector of the data is expected to: • Collect only required data • Use data only for purpose gathered • Collect legal and ethical data only • Preserve confidentiality /integrity of collected personal data • Not misuse the data (e.g., alter image) • Share data only with user consent
What is the purpose of the data collection?	The purpose may be to: • Carry out service for a user • Improve user experience • Understand users • Build up profile of user behaviour • Exploit user data for commercial purposes
How is user consent to data collection given?	User consent may be given in various ways • User informed of purpose of data collection • User consents to use of data • May be hidden in terms and conditions of site
User control	This refers to the ability of the user to control the way that their personal data is being collected/used: • Ability of user to modify their personal data • Ability of user to delete their personal data

and to help, and data scientists have a responsibility to use this power wisely. Data science uses many concepts from the statistics field, and it is essential that both the data and models are valid in order to prevent errors occurring during data analysis.

The consequence of an error in the data analysis or with the analysis method could result in harm to the individual. There are many sources of error such as the sample chosen, which may not be representative of the entire population. Other problems arise with knowledge acquisition by machine learning, where the learning algorithm has used incomplete training data for pattern matching (or other knowledge) recognition. Training data may also be incomplete if the future population differs from the past population.

The data collection involves deciding on the data and attributes to be collected, and often the attributes chosen are limited to what is available, and the data scientist will also need to decide what to do with missing attributes. Often errors arise in data processing tasks such as analysing text information or recognising faces from photos. There may be human errors in the data (e.g., spelling errors or where the data field was misunderstood), and errors may lead to poor results and possible harm to the user. The problem with such errors is that often decisions are made on the basis of data, and individuals are unaware as to whether there is a way to correct errors in the data.

Even with perfect data the conclusions from the analysis may be invalid due to errors in the model, and there are many ways in which the model may be incorrect. Many machine-learning algorithms just estimate parameters to fit a pre-determined model, without knowing whether the model is appropriate or not (e.g., the model may be attempting to fit a linear model to a non-linear reality). This becomes problematic when estimating (or extrapolating) values outside of the given data unless there is confidence in the correctness of the model.

Further, care is required before assigning results to an individual from an analysis of group data, as there may be other explanations (e.g., Simpson's paradox in probability/statistics is where a trend that appears in several groups of data disappears or reverses when these groups are combined). It is important to think about the population that you are studying, and to make sure that you are collecting data on the right population, and whether to segment it into population groups, as well as how best to do the segmentation. In summary, errors may arise due to:

- Human errors in the data
- Bias may be unintentionally introduced
- Training data may not be representative of population
- Data may be presented in a misleading way
- Errors in the model
- Incorrect pre-determined model.

It may seem reasonable to assume that data-driven analysis is fair and neutral, but unfortunately the problem is that humans may unintentionally introduce bias, as they set the boundary conditions. The bias may be through their choice of the model, the use of training data that may not be representative of the population, or the past population may not be representative of the future population, and so on. This may potentially lead to algorithmic decisions that are unfair (e.g., the infamous Amazon hiring algorithm is discussed in Chap. 7), and so the question is how can we be confident that the algorithms are fair and unbiased? Data scientists have a responsibility to ensure that their algorithms are fit for purpose and use the right training data, and as far as practical detect and eliminate unintentional discrimination (individual or target group).

Another problem that may arise is data that is correct but presented in a misleading way. One simple way to do this is to manipulate the scales of the axis to present the results visually in an exaggerated way. Another example is where the results are correct, but presented in an over simplistic manner (e.g., there may be two or more groups in the population with distinct behaviour where one group is the dominant), where the correct aggregate outcome is presented but this is misleading due to the differences within the population, and by suppressing diversity there may be discrimination against the minority group. In other words, the algorithm may use criteria tuned to fit the majority and may be unfair to minorities.

Exploration is the first phase in data analysis, and *a hypothesis may be devised to fit the observed data* (this is the opposite of traditional approaches in statistics where the starting point is the hypothesis, and the data is used to confirm or reject the hypothesis based on the data from the control and target groups, and so this approach needs to be used carefully to ensure the validity of the results).

5.3.2 Problems with Data Science Algorithms

Data science algorithms tend to learn and codify the current state of the world, and it is therefore harder to change the algorithm to reflect the reality of a changing world. The impact of innovative technologies affects the different cohorts and social groups in society in different ways, and there may also be differences between how different groups view privacy. Data scientists tend to be focused on getting the algorithm to perform correctly to do the right processing, and so often they may not consider the wider societal impacts of the technology.

Algorithms may be unfair to individuals in that an individual may be classified as being a member of a group in view of the value of a particular attribute, and so the individual could be typecast due to their perceived membership of the group. Another words, the individual may be assigned opinions or properties of the group, and this means that there is a danger of developing a stereotype view of the individual. Further, it may be difficult for individuals to break out of these stereotypes, as these biases become embedded within the algorithm thereby helping to maintain the status quo.

There are further dangers when predictions are made, as predictions are probabilistic and may be wrong, and only suggest a greater likelihood of occurrence of an event. Predictive techniques have been applied to predictive policing and to the prediction of uprisings (see Chap. 6), but there are dangers of false positives and false negatives.

It is important that the societal consequences of algorithms are fully considered by companies to ensure that the benefits of data science are achieved, and harm to individuals is avoided.

5.4 Data Analytics

Data analytics is the science of handling data collection by computer driven systems, where the goal is to generate insights that will improve decision-making. It involves the overlap of several disciplines such as statistics, information technology and domain knowledge, and is widely used in social media, e-commerce, the Internet of Things, recommendation engines, and gaming.

Data analytics involves the analysis of data to create order and patterns from the data, and it creates information as well as generating insights from the data. This is essential in making informed decisions to meet current and future business needs. There are four types of data analytics (Table 5.3).

Descriptive analysis is a data analysis method that is used to give a summary of what is going on and nothing more. It provides information as to what happened, and it allows the data collected by the system to be used to identify what went wrong. This type of data is often used to summarise large data sets, and to describe a summary of the outcomes to the stakeholders. The most relevant metrics produced include the key performance indicators (KPIs).

Diagnostic analysis is concerned with the analysis of the descriptive metrics to solve problems, to identify what happened, and to understand why something has happened.

Predictive analysis involves predicting what is likely to happen in the future based on data from the past: i.e., it is attempting to predict the future based on actions in the past, and it may involve the use of statistics and modelling to predict future performance, based on current and historical data. Other techniques employed include neural networks, linear and multiple regression analysis, and decision trees.

Prescriptive analysis is used to help business to make better decisions through the analysis of data, and is effective when the organization knows the right questions to ask and responds appropriately to the answers. It often uses AI techniques such as machine learning to process a vast amount of data, to find patterns and to recommend a course of action that will resolve or improve the situation. The recommended course of action is based on past events and outcomes, and the use

Table 5.3 Types of data analytics

Type	Description
Descriptive	These metrics describe what happened in the past and gives a summary of what is going on
Diagnostic	These are concerned with why it happened and involve analysis to determine why something has happened
Predictive	These are concerned with what is likely to happen in the future
Prescriptive	These are concerned with analysis to make better decisions, and it may involve considering several factors to suggest a course of action for the business. It may involve the use of AI techniques such as machine learning, and the goal is to make progress and avoid problems in the future

of machine learning strategies builds upon the predictive analysis of what is likely to happen to recommend a future course of action.

Prescriptive analytics may be used to automate prices based on several factors such as demand, weather, and commodity prices. These algorithms may automatically raise or lower prices at a much faster rate than human intervention, and examples include the automated changing of prices for airline flights and hotels rooms.

Companies may use data analytics to drill down into customer data to determine what they are looking for. This includes understanding the features desired of the product and the price that they are willing to pay, and so data analytics has a role to play in new product design. They may be used by the business to improve customer loyalty and retention, and this may be done by gathering data (e.g., the opinions of customers from social media, email, and phone calls) to ensure that the voice of the customer is heard and acted upon.

Marketing groups often use data analytics to determine how successful their marketing campaign has been, and to make changes where required. The marketing team may use the analytics to run targeted marketing and advertisement campaigns to segmented audiences (i.e., subsets of the population based on their unique characteristics such as demographics, interests, needs and location). Market segmentation is useful in getting to know the customers, and determining what is needed in their market segment, and to determine how best to meet their needs.

Big data analytics may be used for targeted advertisements. For example, Netflix collects data on its customers including their searches and viewing history, and this data provides an insight into the specific interests of the customer, which is then used to send suggestions to the customer on the next movie that they should watch.

Big data analytics involves examining large amounts of data to identify the hidden patterns and correlations, and to give insights to enable the right business decisions to be made. It allows the business to collect as much data as required to understand their customers and to derive important insights.

5.4.1 Business Analytics and Business Intelligence

The effectiveness of management decision-making is influenced by the accuracy and completeness of the information that managers have, with inaccurate or incomplete information leading to poor decisions. Companies often have data that is unstructured or in diverse formats, and such data is generally more difficult to gather and analyse. This has led software firms to offer business intelligence solutions to organizations that wish to make better use of their data, and to optimise the information gathered from the data. There are several software applications designed to unify a company's data and analytics.

Business analytics involves converting business data into useful business information through the use of statistical techniques and advanced software. It includes a set of analytical methods for solving problems and assisting decision-making, especially in the context of vast quantities of data. The combination of analysis with intuition provides useful insights into business organizations, and helps them to achieve their objectives. Many organisations use the principles and practice of business analytics.

Business intelligence (BI) processes all of the data generated by a business, and uses it to generate clear reports (e.g., a dashboard report of the key metrics), as well as the key trends and performance measures that management use in decision making. That is, business intelligence is data analytics with insight that allows managers to make informed decisions, and it may employ data mining, performance benchmarking, process analysis, and descriptive analytics.

5.4.2 Big Data and Data Mining

The term *"Big data"* refers to the large, diverse sets of data that arrives at ever-increasing rates and volumes. It encompasses the volume of data, the velocity or speed at which it is created and collected, and the variety or scope of the data points being covered (these are generally referred to as the three V's of big data). There has been an explosion in the volume of big data with zettabytes[2] (ZB) of data employed globally.

Big data often comes from *data mining*, where data mining involves exploring and analyzing large blocks of data to gather meaningful patterns and trends. Data is gathered and loaded into data warehouses by organizations (i.e., the data is centralized into a single database or program), and then stored either on in-house servers or on the cloud. The user decides how to organize the data, and application software sorts the data accordingly, and the data is presented in an easy to read format such as a graph or report.

The data may be internal or external. It may be structured or unstructured, where structured data is often already managed in the organisation's databases or spreadsheets, and may be numeric and easily formatted. Unstructured data uses data that may be unformatted, and so it does not fall into a predetermined format (i.e., it is free form), and it may come from search engines or from forum discussions on social media.

Big data may be collected in various ways such as from publicly shared comments on social media, or gathered from apps, through questionnaires, product purchases, and so on. Big data is generally stored in databases, and is analysed with software that is designed to handle large and complex data sets.

---

[2] A zettabyte is 1 sextillion bytes $= 2^{70}$ bytes (approximately a billion terabytes or 1000 exabytes or a trillion gigabytes).

5.4.3 Data Analytics for Social Media

Data analytics is used by social media companies to provide a quantitative insight into human behaviour on a social media website, and enables the business to understand its audience better, to improve the user experience, and to create content that will be of interest to them. Data analytics consist of a collection of data that says something about the social media conversation, and it involves the collection, monitoring, analysis, summarisation, and graphs to visualise insight into the behaviour of users.

Another words, *data analytics* involves learning to read a social media community through data, and the interpretations of the quantifiable data (or metrics) gives information on the activities, events, and conversations carried out by the users. This includes what users like when they are online and also includes other important information such as their opinions and emotions that need to be gathered through *social listening*. Social listening involves monitoring keywords and mentions in social media conversations in the target audience and industry, to understand and analyse what the audience is saying about the business and allows the business to engage with its audience.

A social media campaign is a marketing tool where a company or social media company uses social media platforms to interact with their customers/users. The design of a social media campaign is often an iterative process, with the fist step being to determine the objectives of the campaign and designing the campaign to meet the requirements. The goals may be to increase the number of users or to build a brand, and data analytics combined with social listening help in understanding how people are interacting, as well as what they are interacting about, and how successful the interactions has been. The campaign may be refined to meet its goals and the cycle repeats.

The key performance indicators (KPI) may include increased followers/subscribers or an increase in the content shared, and so on. Elementary data such as the number of likes, the number of followers, the number of times a video is played Youtube, and so on are gathered to obtained a quantified understanding of a conversation.

Facebook and Twitter maintain a comprehensive set of measurements for data analytics, with Facebook maintaining metrics such as the number of page views and the number of likes and reach of posts (i.e., the number of people who saw posts at least once). Twitter includes a dashboard view to summarise how successful tweet activity has been, as well as the interests and locations of the user's followers. Social media data contains a rich collection of human emotions with the social listening data including user opinions and emotions.

5.4.4 Mathematics Used in Data Analytics

The main mathematics used in data science and analytics include (Table 5.4).

Table 5.4 Mathematics in data analytics

Type	Description
Probability	An introduction to some of the concepts in probability theory such as basic probability, expectation, conditional probability, Baye's Theorem, and probability density functions
Statistics	Statistics is a vast area and includes descriptive statistics; measures of central tendency such as the mean, mode and median, variance and covariance, and correlation and regression
Linear Algebra	This includes topics such as matrix theory and Gaussian elimination as well as basic algebra
Calculus	This includes the study of differentiation and integration and includes topics such as limits, continuity, rules of differentiation, Taylor's series, and area and volume as discussed

Probability theory, statistics, linear algebra and calculus are described in more detail in [3]. Other areas of mathematics that may arise in Data Analytics include Discrete Mathematics and Graph Theory and Operations Research (see [4]).

5.5 Data Privacy

In Greek mythology there was a giant called Argus Panoptes, who was an all-seeing giant with one hundred eyes looking in every direction, and he would always have some eyes open even when asleep. That is, he was always watching and monitoring the world around him, and so he was the perfect guardian. He was later slain by Hermes (the messenger of the gods), who disguised himself as a shepherd and put all of Argus's eyes to sleep, and then slew him.

Jeremy Bentham (the founder of Utilitarianism which was discussed in Chap. 2) designed a circular prison in the eighteenth century termed the Panopticon, where a single guard in the centre of the complex could observe all of the prisoners. His idea was that although individual prisoners did not know if they were being watched or not at a given time instant (as this depended on the direction that the guard was facing), that they would behave as if they were being watched, and so they would behave all of the time (Fig. 5.1).

The modern version of the Panopticon is a set of security cameras that is watching people, or websites that are monitoring user behaviour, or the entire Internet, which is working out everything that we are doing by watching us. The question is whether we as individuals should be concerned about this, and whether it matters if we as individuals are doing nothing wrong. Some have argued that everyone would be completely honest due to zero privacy, and where everyone could know what everyone else is doing. Others respond by saying that privacy is a basic human right, and that it is needed for freedom to be exercised in society.

The "*Right to have privacy*" was an influential legal article that was written by Louis Brandeis and Samuel Warren, and published in the Harvard Law Review

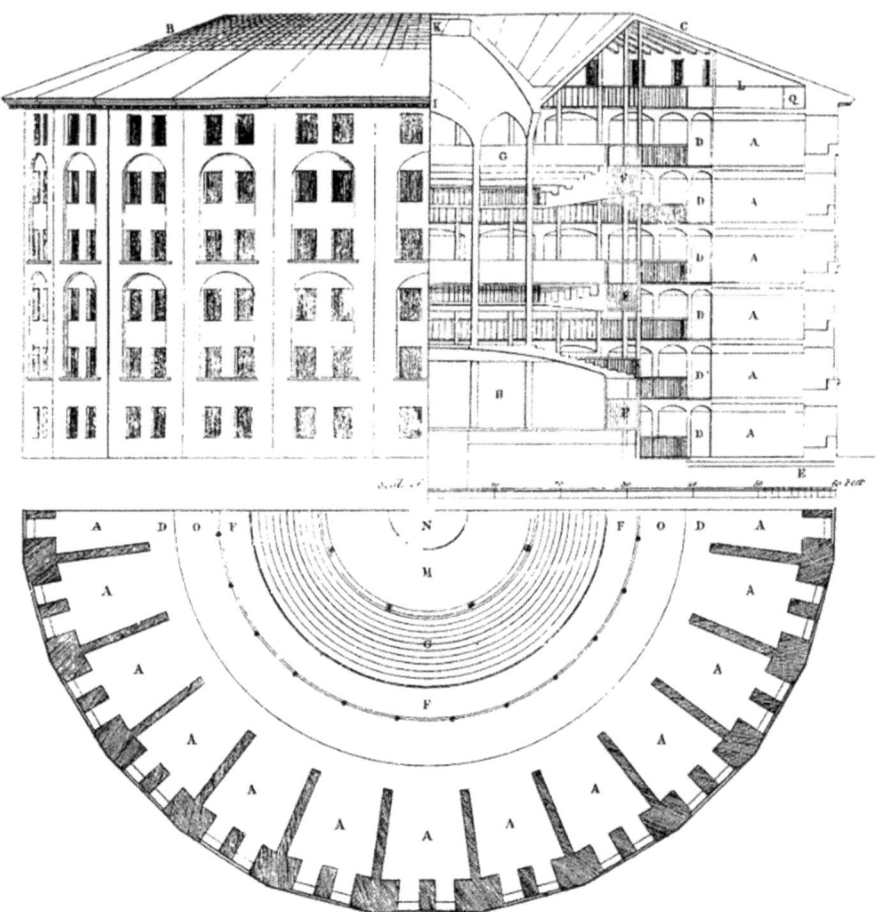

Fig. 5.1 Bentham's Panopticon prison

in 1890 [1]. The article advocates for privacy as *"the right to be left alone"*. William Prosser wrote an article in the Californian Law Review in 1960 in which he outlined four types of privacy torts [2]:

- Intrusion upon seclusion
- Public disclosure of private facts
- Publishing objectionable, false information
- Misappropriation of name or likeness.

There has traditionally been a difference between rural and urban living, where in a small town people know everything about every other person in the town, and there is essentially very little privacy from all the gossip (*pueblo pequeño infierno grande*). In a larger city, people are anonymous and nobody knows or cares about what others are doing, and so there is a greater sense of privacy. Further, an individual living in a small town has a choice of moving to a new town for a fresh start, or waiting in the town for the community to forget a particular event, whereas in a large city this problem is a lot less relevant due to the anonymous nature of city living.

There are some parallels of the Internet being like the small village, except that the relationship is asymmetric. Another words, in a small town each individual knows as much about another as vice versa (i.e., it is a symmetric relationship), whereas the relationship is asymmetric for the Internet. This makes it a very unequal relationship, with one party gathering lots of information and building up a profile about all other parties, and using that information for commercial purposes. The other parties are not actively gathering information, and have a very limited picture of what is going on with all the data that is gathered.

Further, while events and information may be forgotten in a village over time this does not happen with the Internet: i.e., it is very difficult to forget things on the Internet with web pages surviving forever in some archive even if taken down. Another issue is that once a page is put up many copies of it are made, and so even if the page is taken down there may still be many copies remaining elsewhere, and so there is no way of really deleting something once it has been published on the web. This could create major problems for individuals who pose indiscreet content on line, as that content may live on in perpetuity.[3]

People need an understanding of how their personal information and data is collected, shared and used across the many computer platforms that they use, and the extent to which they have control over their personal information. It is important to understand the nature of privacy, the problems and risks that exist, as well as privacy laws that are available to protect individuals from its abuse.

De-identification is the removal of identifiable information from data, and includes the removal of fields (or attributes) such as name, address and phone number so that no personally identifiable attributes remain in the dataset. This means that the identity of the person is not immediately identifiable, and so it provides some safeguards to the individual. However, it is possible that the individual's identity may be determined from the other retained fields, and this means that care must be taken if public records are to be released. That is, it may still be possible (even if de-identification has taken place) to identity individuals. Anonymity is limited or virtually impossible given the extent of public and private information that is available about individuals, and as facial recognition technology has also improved in a major way the identification of individuals from the images of their faces may be rapidly done with technology.

---

[3] It is almost impossible to delete oneself from the Internet. The user could delete some of their personal information such as their social media accounts or restrict public access to these, and also delete any online accounts that are not being used. It is possible to turn on auto-delete on Google to delete your data at regular intervals. Personal information can be deleted from blogs and any apps that are not being used may be deleted.

5.5.1 What is Privacy?

There are various definitions of privacy such as the right to be left alone, for secret or intimate information to be kept secure from others, and for *control over one's personal information* where individuals are able to decide what information will be shared, when it will be shared, and how it will be shared with others.

That is, users should be in control of how their data is used, and most user agreements are "all-or-nothing" in the sense that a user must give up control of their data to use the application. This essentially means that the user is coerced to accept the terms and conditions of the website with the result that the user has no control over how their personal data is used. That is, a user must click acceptance of the terms and conditions in order to use the services of a web application. Clearly, users would be happier and feel that they are in control if they were offered graduated choices by the vendor, to allow them to make tradeoffs, and to choose a level of privacy that they are comfortable with.

Privacy has become quite topical with the rise of social media, the Internet of Things and Artificial Intelligence. However, privacy concerns are not a new phenomena, and they initially grew out of the development of early technologies such as the first cameras, microphones and telephones, where indiscreet or unauthorised images or recordings could be made leading to concerns of an invasion of privacy by a prying media or others (Fig. 5.2).

The early concerns over privacy were with maintaining the security and confidentiality of a message, and so this led to some people and groups to communicate with each other using ciphers. For example, Julius Caesar communicated important messages using an alphabetic cipher during his campaign in Gaul in the first century B.C., and the emperor Augustus used a similar approach for communication. Further, some of the leaders during the American War of Independence used codes and pseudonyms to protect their identity during sensitive communication.

Societies vary in terms of their political systems, with democracies offering a peaceful way of replacing an unpopular government, whereas totalitarian states are often ruthless in their control of the population. Some autocratic societies run by dictators employ a culture of surveillance on the population, and this may include identifying individuals who pose a potential threat to the regime, and removing such individuals from society either by imprisonment or political assassination.[4] Often, these societies are characterised by mass surveillance of individuals, police searches and seizure of private property, police brutality, and so on. In democratic societies there are usually laws to protect the citizen against unreasonable police searches and behaviour.

The importance of privacy in the information technology field became apparent in the early 1970s with the introduction of databases. These could hold private

---

[4] Several oligarchs and critics of Putin and his regime in Russia and the 2022 invasion of Ukraine have died in mysterious circumstances. Some have died by hanging in their own homes, some have fallen out of windows of hotels or hospitals, some have fallen down stairs, some have fallen overboard from their yachts, and some have died in plane crashes.

Fig. 5.2 Cardinals eavesdropping in the Vatican

information about individuals, and there was a need for a set of rules to protect how information should be collected and used. This led to the development of a set of *fair information processing principles* (FIPs) by the US Secretary's Advisory Committee on Automated Personal Data Systems published in their 1973 report on Records, Computers and the Rights of Citizens [5]. This led to the Fair Information Practice Principles Act in 1974, and this act became the basis on which data collection is governed in the United States. The report outlined several principles such as (Table 5.5).

Computing technology has evolved in a major way from the mainframes and databases of the early 1970s, with leading edge technologies such as smart phones, Social Media, the Internet of Things, and Artificial Intelligence. It is reasonable to ask what privacy means in the modern digital world and whether there is privacy anymore? Users of social media share large parts of their lives with a massive on-line audience as well as with large corporations, and social media companies gather lots of data about its users that may be used to determine patterns, and to generate profiles that may be targeted to advertisers. So much data is being collected about individuals, and the question is where does it go? Who controls it? Are companies adequately managing risks of data breaches? What happens when data privacy is breached or data is not secured properly? Is there transparency? Is user data encrypted? Is confidentiality and authenticity maintained?

Table 5.5 Principles of data collection

Principle	Description
Transparency of collection and storage of information	The organisation that is collecting personal data must be doing so openly (i.e., it is not secretly or covertly collecting data)
Accessibility of personal information	An individual must be able to access any data that the organisation has about him/her
Purpose limitations (consent)	There must be a way for an individual to prevent information that was gathered for one purpose to be used for another purpose without their consent
Correction of personal data	There must be a way for an individual to correct or amend information about him/her
Personal data safeguards and accountability	Any organisation that is creating, maintaining, or disseminating personal data must ensure its reliability for the identified use, and take reasonable precautions to prevent against any misuse of the data

The *Internet of Things* (IoT) is not a single technology as such, and instead it is a collection of devices, sensors and services that capture data to monitor and control the world around them. It refers to interconnected technology that is now an integral part of modern society, where computation and data communication are embedded in the environment. It allows everyday devices to connect to other devices or people over the Internet.

The level of interconnectivity and data gathered with IoT means that security and privacy have become important concerns, and it is essential to control both the devices and the data. Data should only be gathered with user consent, and there are risks of hacking or eavesdropping.

There has been a major growth in AI technology in recent years, and AI has been applied to self-driving cars, facial recognition, machine translation and so on. Facial recognition technology may be used to unlock phones to authenticate identity, and it has also been applied to read facial expression during job interviews, as well as following the movement of individuals. A vehicle may contain several on-board computers for processing various vehicle controls as well as for entertainment systems. Vehicles that connect to the Internet are potentially at risk of being hacked, where a hacker may potentially commandeer vehicle controls such as steering and the brakes.

It is often unclear who is collecting the personal information, the type of information that they are collecting, what is being done with the data, and whom the data is being shared with. The theft of personal information results in others having information about a particular individual, and they may be in a position to use that information against the individual. This means that the perpetrators have power over the individual, since they may use the stolen personal information about the individual to commit a crime.

5.5.2 Social Media

Social media involves the use of computer technology for the creation and exchange of user-generated content. These web-based technologies allow users to discuss and modify the created content, and it has led to major changes in communication between individuals, communities and organisations.

Social media is designed to have the individual share personal information while they are on the site, and with every disclosure (or post) the individual reveals a little bit more about himself or herself. It is very easy to post photos and information on social media sites such as Facebook or Twitter, and *social media is designed in such a way that it is addictive and poses risks to the privacy of an individual* (Fig. 5.3).

There is a danger that both social media companies and other users could harm the individual's privacy. The harm from other users may arise when a piece of the user's information is shared with the wrong audience, and this later leads to problems for the user. There are two distinct audiences for the individual's information namely other users and the platform itself. The social media platform maintains a vast quantity of electronic information consisting of immense databases, which can collect a vast amount of data on the individual and others.

There is a power imbalance between the platform and the user, with the platform designed to have the individual share as much as possible, and other users may potentially pose risks to the user with their social interaction. An individual's information may be viewed by friends, family, employer, work colleagues and nameless others, and so everyone in the individual's network as well as others could be an unwanted audience.

Fig. 5.3 Young peoples on smart phones and social media. Public Domain

Table 5.6 Threats in social media

Threat	Description
Manufactured disclosures	This refers to how a social media site gets people to disclose more and more information, and this is similar in a way to surveillance. Traditional surveillance involves watching people to learn something about them, whereas modern surveillance as in social media has less to do with this, and it generally involves getting people to disclose something more about themselves and so, in effect, to learn something new about the person
Extracting consent	This refers to how a social media site obtains consent from its users on the various practices employed on the site. A user must click acceptance of the associated terms and conditions to use the site, and often this involves accepting invasive privacy practices described in a long, dense and largely unreadable terms of use document. The social media site may also request access to the camera, location and address book of the individual. Often, users just accept the terms and conditions and permission requests because they are so worn down with so many requests from different apps, and they have no choice but to accept the terms of use and invasive practices in order to use the site
Overexposure	Social media sites constantly introduce new features to make user data more visible, more searchable, and more complete to others and may result in *over exposure* of the information
Faithless friends	This refers to when information that has been shared in the individual's network is shared more widely by one of the "friends" of the individual. This may lead to embarrassment or harm to the individual
Online harassment	This is where repeated insults or bullying of an individual takes place on line, which may even include threats of violence, posting of indiscreet images and even revenge porn

Users often may not realise the full extent of their audience when they post on the site, as people who are authorised in an individual's network may not be the desired recipients of certain posts (disclosures). It is difficult to delete online messages, and destructive posts may last long after an incident. Therefore, it is very much in the interests of users to keep their posts discreet, as both friends and outsiders of their social media network could pose risks to their privacy. There are several potential threats on social media sites including (Table 5.6).

We will discuss the ethics of social media in more detail in Chap. 6.

5.5.3 Internet of Things

The Internet of Things is a collection of devices, sensors and services that capture data to monitor and control the world around them, and these include devices and sensors such as cars, clothing, fridges, fitness monitors, and so on. An individual may be continuously connected to multiple home devices with sensors (e.g., microphones and cameras), and connection and access to these devices increases the risk to data security (Fig. 5.4).

5.5 Data Privacy

Fig. 5.4 Fitbit surge. Smart-watch activity tracker. Creative Commons

The fact that there are many devices with sensors connected to the Internet means that there are, in effect, more eyes watching the individual and gathering data about her, and there are also more points of failure. This means that IoT poses greater risks to the safety of individuals than when using basic computers, and the risks include:

- Security risks
- Privacy risks.

These devices consist of both hardware and software and so there are now two points of failure: i.e., hardware failure and software failure. Hardware is generally more reliable than software, and hardware failures tend to be as a result of components wearing out over time. Software failures are often as a result of design issues, and software often requires regular updates to correct problems or to deal with security vulnerabilities. The fact that these devices are connected to the Internet means that software upgrades are possible, but being connected to the Internet means that the device may be targeted by hackers in a similar way to which a computer is hacked.

Further, these Internet devices contain sensors that gather a lot of personal data about the individual, and they collect, use and share this data, and so the IoT devices pose similar data security risks as laptops or smart phones. Many IoT devices are inexpensive and have serious security vulnerabilities, with some Internet devices failing to encrypt data when transmitting data or images to the cloud. This means that that an eavesdropper could intercept this Internet traffic, and cause harm to the individual.

IoT has serious implications for privacy in that these devices can produce granular personal data such as when the individual is at home, what the individual eats, and so on. They gather a lot of personal data and the data may be shared with other devices or platforms thereby posing risks to the privacy of the individual.

5.5.4 AI and Facial Recognition

There has been a major growth in the AI field in recent years, and facial recognition is an AI technology that offers the ability to unlock phones to authenticate identity, and so it may be used to protect the individual. Facial recognition has advanced in sophistication to allow individuals be recognised at demonstrations and street protests, and this means that some authoritarian governments may potentially use this technology as a tool for authoritarian control. That is, surveillance technology combined with facial recognition could be oppressive to individuals and society, and lead to a totalitarian state.

A society that adopts a paradigm of constant surveillance, where individuals are living in a world with technology constantly monitoring their daily activities, learning to recognise patterns, and drawing inferences is moving towards totalitarianism.[5] Facial recognition is a potentially dangerous technology that may challenge civil liberties, and it could severely impact marginalized groups in society. Facial recognition has become popular since:

- Faces are central to identity
- Faces are hard to hide
- Existing face and name databases
- Facial recognition is widespread.

Facial recognition is a biometric technology that analyses visual data from social media and other sources, and is capable of detecting facial features and to essentially reduce each face to a mathematical equation using factors such as the distance between the individual's eyes, the width of the nose, and so on, and the patterns in the visual data are compared to patterns in facial recognition databases to confirm identity.

Some companies have applied facial recognition technology to read facial expression during job interviews, and this provides a mechanism for the company to obtain data that they may not otherwise receive. Deep fakes are an AI technology that allows convincing images and videos to be created of individuals doing things that they never did or said, and this disruptive technology has been applied to misrepresent individuals in a variety of ways. An individual may be seen to make false or even preposterous claims, and this is achieved from content taken from social media and other media that is then manipulated and edited in various ways to achieve the desire effect. This technology could potentially show an individual committing a crime or present the individual in a very negative way. The technology has at this time been mainly been applied to humour as in political

---

[5] Orwell's famous 1949 novel "Nineteen Eighty Four" is based on Stalinist Russia and Nazi Germany, and describes a totalitarian state where the Ministry of Truth engages in omnipresent government surveillance and constant propaganda to repress individuality and independent thinking.

satire, but as the technology improves it may become difficult to distinguish the real from the fake with potential serious consequences for society.

5.5.5 Privacy and the Law

Data collection laws focus on how data is collected, used and shared, and data protection includes the right to information self-determination.[6] The web is full of privacy policies that specify what type of personal data will be collected, how it will be processed and used, how it is shared, and what can be done about it. Individuals may take a lawsuit against another for a tort, for example, when someone pries or stalks them, or publishes a defamatory article, or violates their privacy. There are three main areas that impact upon an individual's privacy namely:

- The Media
- Surveillance
- Personal Data.

Media laws protect an individual against intrusion, where another party may be held liable for the invasion of the individual's privacy (e.g., phone tapping, snooping, examining a person's bank account, and so on). The tort of the public disclosure of private facts is part of the legal system in many states, and its goal is to prevent the public disclosure of private facts concerning the private life of an individual, where the matter is not of legitimate concern to the public. That is, others are prevented from widely spreading private facts such as the individual's face or identity for their own benefit, and there are slander and libel laws to protect an individual's good name and reputation, and to prevent defamation of character.

There are laws and rights to regulate surveillance with search warrants required in most countries to search the home of a private individual and to seize personal property. Warrants are generally required to obtain personal electronic records held by telecommunication companies (e.g., the calls made and received as well as meta data such as the date, time and duration of the call as well as geo-location data), and warrants may be required to obtain records held by Internet technology companies (e.g., emails, web sites visited, searches, and other electronic messages).

Countries vary in their laws for the protection of security and privacy, but in general the security and privacy commitments made by a company in their policies should be fully implemented. Companies should be held accountable for any security breaches that occur that lead to privacy being compromised, and they may be held liable for any losses suffered by individuals as a result of the breach.

---

[6] Information self-determination refers to the freedom that individuals have to decide what happens to their personal data as well as the purpose of data collection.

Further, people must not be misled about the functionality of a website or mobile app that places their security or privacy at risk, and users must give their consent to any changes to the privacy policy that would allow for the collection of additional personal data, and users must be informed about the extensiveness of tracking and data collection.

5.5.6 EU GDPR Privacy Law

Europe has been active in the development of data protection regulation, and the European General Data Protection Regulation (EU GDPR 2016/679) is a comprehensive data protection framework that became operational in 2018. The importance of both privacy and data protection has been recognised in Europe, and these are regarded as fundamental human rights in the EU. The goal is to give individuals control over their personal data, and it has had a huge impact on privacy laws of other countries around the world, with other countries using it to develop similar laws (e.g., Japan and the state of California in the US).

GDPR also addresses the transfer of personal data outside of the EU, and it prohibits the transfer of personal data outside of the EU to countries that do not provide an equivalent or adequate data protection framework as GDPR. GDPR consists of a data governance framework that attempts to place privacy on a par with other laws. It creates protections that follow the data, and it places responsibilities on companies to manage privacy and information (Fig. 5.5).

GDPR applies whenever personal data is processed, and it starts from the presumption that the processing of the personal data is illegitimate. This means that companies carry the burden of legitimising their actions, and they must be able to show that they have a legitimate basis for processing data. That is, they must be able to show that they have the consent of the data subject, or that the processing is necessary as a result of the contract that exists between them and the data subject, or where they have a legitimate interest, and where the interest of the data controller prevails over that of the data subject. The company must be able to demonstrate adherence to the fair information practice below:

Fig. 5.5 EU GDPR 2016/679

5.5 Data Privacy

- Standards for data quality
- Standards for transparency
- Special protections for sensitive data
- Standards of enforcement.

This means that data must be obtained legitimately and used for the purpose for which it was acquired, and there must be openness and transparency so that individuals will know how their data will be used. There should be special protections for sensitive data with the ability to opt in for consent (e.g., race, sexual orientation, and so on), and there must be standards for enforcement to ensure compliance to the standards. The *Data Privacy Impact Assessment* (DPIA) is part of GDPR, and is needed if the processing of personal information is likely to result in a high risk to the rights and freedoms of individuals.

The standard for informed consent is very high which means that it is freely given and informed. GDPR also gives very strong data subject rights, including the right to access data, data portability, the right to rectify data, the right to erase data, the right to object to processing, and the right to restrict processing. These rights provide a tool for data subjects to exercise control over their personal information.

The EU Data Protection Commissioner (DPC) is responsible for monitoring the application of GDPR in the EU, and for upholding the fundamental rights of EU citizens in protecting their personal data. Several of the large technology companies (e.g., Amazon, Facebook and Google) have been subject to large fines in the EU for violations of GDPR.

5.5.7 EU Digital Services Act

The European Digital Service Act (DSA) became part of European Law in 2024, and it is applicable to all member states in the European Union. The goal of this legislation is to protect digital space against illegal content as well as protecting the fundamental rights of users, and it is applicable to social media companies and the online marketplace. Another words, DSA is a set of rules and regulations to provide a safer online digital space, and protect the fundamental rights of users of digital services in the European Union.

Platforms will be held accountable (in Europe) for their role in disseminating harmful content, and the legislation lays down obligations on platforms and control measures. These include:

- Provide greater transparency on their services
- Obligations on market place to combat sale of illegal products/services
- Measures to combat illegal content and obligations on platforms to respond quickly
- Protection of minors by prohibiting targeted advertisements
- Limits on advertisement

- Stricter rules for very large online platforms (VLOPs) and search engineers (VLOSEs)
- Improve control for users of their service.

DSA is applicable to very large online platforms (VLOPs) and very large online search engines (VLOSEs), as well as online platforms, hosting services, and Intermediary Service Providers (ISP). Member states have the power to supervise and enforce the DSA in relation to ISPs that have their main establishment in that member state, and each member state designates a Digital Services Coordinator (DSC) to supervise and enforce the DSA in that state.

The European Commission has the power to supervise and enforce the Digital Service Act in relation to very large platforms or search engines that are classified as VLOPs and VLOSEs, and these are subject to additional stricter rules.

5.6 Security of Data

The privacy of personal information may be violated if there is weak system security, and so good cybersecurity is a prerequisite for the protection of personal data. Poor security may lead to the theft of personal information, and the security of the system refers to its ability to protect itself from attack. A secure system will preserve important characteristics of the data such as:

- Confidentiality
- Integrity
- Availability.

Confidentiality means that the information may be viewed and accessed only by those authorised, and encryption may be employed to ensure that the unauthorised access of information is meaningless to anyone other than the intended parties. Other approaches include access controls where only those with the appropriate access privileges may access the data. *Integrity* means that the data may only be modified by those authorised to do so, and *availability* refers to the fact that the system and its data are available for use at all times (i.e., it is not subject to a denial of service attack). Attacks may be both internal or external to the company and may lead to:

- Unauthorised data access and usage
- Unauthorised data theft and deletion
- Unauthorised data manipulation.

The system needs to be designed for security, and security loopholes may be introduced in the development of the system, and so care needs to be taken to prevent these. A risk profile of the system may be determined from risk analysis to determine threats and vulnerabilities as well as their probability and impacts. The high-risk areas lead to the security requirements including the required security measures and supporting technologies. There is a trade off between security risks and the cost of security measures, and this is a continuous process due to continued changes in technology. A comprehensive security system requires a range of measures such as:

- Preventive measures
- Detective measures
- Administrative measures.

Preventive measures are used to stop unauthorised attacks from occurring before they succeed and do any harm; *detective measures* are used to discover any unauthorised attacks that may be on-going or completed; and *administrative measures* which are used to clarify processes, rules and standards within an organisation. Organisational and administrative measures are as important as technical measures in securing a system.

The administrative measures include identifying the technical measures, actions and enforcement mechanisms that are needed, and defining the responsibilities for carrying out the measures. The policies and procedures as defined in the Information Security Management System of ISO/IEC 27,001 provide guidance on what should be implemented in security management.

Preventive measures may include the use of encryption for communication and stored data, so that the data is meaningless to anyone who is not authorised to see it. Preventive measures also include access control mechanisms for *authentication* to verify if the person is who she claims to be, and a range of measures such as user id and password, smart card and biometric data may be used for verification. The next step is to ensure that those authenticated have the appropriate level of *authorisation* to access the data, and an authorisation matrix that includes roles and the level of access for each role may be used for this.

The goal of detective measures is to monitor whether a system is actually secure, and to detect attacks. Security audits may be conducted to verify that the planned security measures have been implemented. Penetration testing is a way to find and remove security weaknesses, and it involves experts (e.g., white hat hackers) playing the role of attackers trying to find vulnerabilities in the system, and taking actions to address any identified weakness in the system. Security is discussed in more detail in Chap. 12.

5.7 Review Questions

1. What is data science?
2. What is the role of the data scientist?

3. What is privacy? Why is it important?
4. What are the main sources of personal data collected on line?
5. What are the main risks to an individual using social media?
6. What are the main risks to an individual using a fitness device (as part of the Internet of Things)?
7. What are the main risks with AI facial recognition technology?
8. Explain the importance of the EU GDPR law.

5.8 Summary

The ethics of data science aims to investigate what is fair and ethical in data science, and what should or should not be done with data. A fundamental principle of ethics in data science refers to informed consent, and this is where the individual is informed about the purpose of data collection and gives their consent voluntarily. This consent is often given by individuals accepting the terms and conditions before they may use software applications, and these state that data will be collected, processed and shared.

Companies collect lots of personal data about individuals from their use of computer resources, and this data is processed to build up revealing profiles of the users that may be sold on to advertisers. Mass surveillance may be conducted by governments on its citizens, with face recognition software allowing citizens to be recognised at demonstrations or other mass assemblies.

Privacy remains important in the information age, and it is the way in which we separate ourselves from other people, and it is the right to be left alone. Modern technology allows the location of the user to be tracked, and poses challenges to privacy. However, the European GDPR law has become an important protector of privacy and personal data, and both European and other countries have adapted it.

References

1. L. Brandeis, S. Warren, *The Right to Have Privacy* (Harvard Law Review, 1890)
2. W. Prosser, Privacy. Californ. Law Rev. (1960)
3. G. O' Regan, *Mathematics in Computing*, 2nd edn. (Springer, Berlin, 2020)
4. G. O' Regan, *Guide to Discrete Mathematics*, 2nd edn. (Springer, Berlin, 2021)
5. US Secretary's Advisory Committee on Automated Personal Data Systems, in *Records, Computers and the Rights of Citizens*. https://aspe.hhs.gov/report/records-computers-and-rights-citizens

Ethical Social Media

Abstract

This chapter discusses ethical social media, and we discuss the Facebook revolution and its impact during the Arab spring as well as social media campaigns. We discuss the Cambridge Analytica affair and its impact in influencing voters in the 2016 election in the United States.

Keywords

Facebook Revolution · Arab Spring · Social Media Campaigns · Cambridge Analytica Affair · Moderating Content · Fake News and Disinformation · Ethical Social Media · Data Analytics · Tweet

6.1 Introduction

Social media involves the use of computer technology for the creation and exchange of user-generated content. These web-based technologies allow users to discuss and modify user created content, and it has led to major changes in communication between individuals, communities and organisations. It plays a major role in connecting people and enabling communication, and it has a role to play in social engagement on civic matters.

Social media provides a way for like-minded people to gather together, and is an effective way to mobilise protest. Politicians use social media during election campaigns as a way to communicate their message and policies to a wider audience, and to engage with their constituents on a day-to-day basis during sessions of parliament. Social media enables citizens to play a role in shaping the world around them (e.g., on climate change) and to make a difference. Corporations may use social media as a way to get closer to their customers, so that they may understand what customers like and dislike about their products and services, with the goal of improving to serve them better.

On line communities emerged out of the first Internet chat rooms, and one of the earliest online communities was that developed on the PLATO system at the University of Illinois in the early 1970s. The PLATO system pioneered online forums, message boards, email, chat rooms, and multiplayer games, and it was designed for computer-based education. Usenet newsgroups and message boards were developed for the Internet in the early 1980s, and CompuServe introduced chat rooms around the same time. The invention of the World Wide Web in the late 1980s by Tim Berners Lee led to further innovations such as AOL, Yahoo and MSN messenger in the mid 1990s. Social media took off in a major way from the early twenty-first century.

There are several popular social media sites such as Facebook (a social media site that allows users to communicate with family and friends in an informal way), Twitter/X (a social media site where users may send short messages called tweets to their followers), LinkedIn (for more formal communication to a user's professional network), and Instagram (a photo and video sharing social media site). Often, there are boundaries between these sites, as LinkedIn and Twitter are often used for professional activities, whereas Facebook and Instagram are used mainly used for social and personal activities. This means that there may be differences in the formality of communication on these platforms.

Social media users may be active or passive, with *active users* actively contributing content to the site, and *passive users* often just observing or following events. Social media platforms are generally free to use, but they gather a lot of data about users, and so this has led to concerns over privacy and surveillance. That is, *the users are essentially the product*, and the social media platform gathers data about them thereby building up a profile of the users that may be sold on to advertisers and other interested commercial parties. Social media provides benefit to the public as well as to commercial organisations.

Facebook has facilitated a cultural shift in human communication, and it enables users to keep in touch with friends and family around the world, and to share their opinions on what is happening in the world. Users may upload photos and videos; express opinions and ideas; and exchange messages. Facebook allows the user's community of friends to be kept up to date on important events that the user wishes to share. It has become an important channel for educated young people to discuss their aspirations for the future, as well as their grievances with society and the state, as well as being an effective tool for protest.

Twitter has become an effective way to communicate the latest news, and its effectiveness as a communication tool increases as the number of a person's followers grows. It allows a person or organisation to determine what people are saying about it, including their positive or negative experiences. This allows direct interaction with the followers, and so it is a powerful way to engage the audience and to make people feel heard.

LinkedIn has become an important on-line platform for professionals, and it allows users to create a profile that summarises their education, skills, experience and previous employment. It is a useful way for professionals to build connections with industrial contacts, and to become aware of new opportunities.

6.2 The Facebook Revolution

Facebook is the leading social networking site (SNS) in the world, and its mission is to make the world more open and connected. It is a powerful communication tool that enables users to send and receive messages, and to keep in touch with friends and family around the world. It allows users to create and share information such as their opinions on what is happening in the world around them. Users may upload photos and videos, express opinions and ideas, exchange messages, and collaborate. A Facebook community is a key concept, and it consists of a group of like-minded individuals that are interacting for a common goal. Facebook is very popular with advertisers as it allows them to easily reach a large target audience.

Mark Zuckerberg founded the company in 2004 while he was a student studying psychology at Harvard University. Zuckerberg was interested in programming, and he had already developed several social networking websites for his fellow students including *Facemash* which could be used to rate the attractiveness of a person, and *Coursematch* which allowed students to view people taking their degree (Fig. 6.1).

Zuckerberg launched "*The Facebook*" (thefacebook.com) at Harvard in February 2004, and over a thousand Harvard students had registered on the site within the first 24 h. Over half of the Harvard student population had a profile on Facebook within the first month. The membership of the site was initially restricted to students at Harvard, then to students at the other universities in Boston, and then to students at the other universities in the United States. Its membership was extended to international universities from 2005.

Fig. 6.1 Mark Zuckerberg

The use of Facebook was extended beyond universities to anyone with an email address from 2006, and the number of registered users began to increase exponentially. The number of registered users reached 100 million in 2008, 500 million in 2010; it exceeded 1 billion in 2012, 2 billion in 2017, and close to 3 billion in 2022. It is now one of the most popular web sites in the world.

Facebook's business model is quite distinct from that of a traditional business in that it does not manufacture or sell any products. Instead, it earns its revenue mainly from advertisements, and its business model is based on advertisement revenue, with advertisements targeted to its roughly 3 billion users based on their specific interests. Another words, Facebook is essentially selling its users to advertisers (i.e., the users are the product), and the users do all the work with Facebook monitoring their activities on the site. Facebook collects data about the users (e.g., their age, gender, location, education, work history and interests), and classifies and categories the users, so that advertisers may target advertisements to them. This ensures that the advertisements are targeted to the right audience, and will potentially be of interest to the targeted audience.

6.2.1 The Arab Spring

The Arab Spring refers to a series of anti-government protests from 2010 to 2012 that took place in various parts of the Arab world including Tunisia, Libya, Egypt, Syria, and Yemen, where rulers were forcibly removed from power or major uprisings took place (Fig. 6.2).

The Arab world had been ruled peacefully for several hundred years by the Ottoman Empire, and the Anglo-French *Sykes Picot agreement* defined how the Middle East would be divided up between Britain and France after the First World War. This agreement led to the creation of several new nation states, where there had previous been no state, and this included the creation of new countries such as Iraq, Syria, Transjordan, Lebanon and Palestine The division of the Middle East between Britain and France involved drawing lots of straight lines without seriously thinking through the consequences of the borders.

The result was that Arab tribes were often divided (the tribe is a fundamental part of Arab identity and determines the status of a person in Arabic society), and it led to states where there were significant ethnic or religious minorities. For example, Iraq had a large population of Shia Muslims in the south, a large population of Kurds (who were non-Arabic Sunni Muslims in the north), and Sunni Muslims in the middle. The split between Shias and Sunnis arose in relation to succession and leadership of the Muslim world after the death of the Prophet, and the deep disagreements between Shias and Sunnis are mainly political rather than theological. Further, for Iraq under the Ba'ath Party and later Saddham Hussein the Sunnis dominated the country, and the Kurds and the Shias were crushed into submission, and so there were deep sectarian divisions and limited social cohesion in the state.

Fig. 6.2 Arab Spring in Middle East

There were similar problems with the creation of many of the other countries in the Middle East, although Jordan (formerly called Transjordan) has been the most stable of the new Arab states created after the First World War. Jordan is mainly a homogenous society that is mainly Sunni, but some instability has arisen in recent years from the large number of Syrian refugees that arrived in the country following the collapse of Syria. The situation in Palestine was even more complex, as the 1917 *Balfour declaration* had indicated that the British government would view with favour the creation of a national home for the Jewish people in Palestine. The declaration emphasised that such a home must not prejudice the civil and religious rights of others living in Palestine, but it did not (at the time of the declaration) give a legal commitment to the creation of a Jewish state in Palestine.

However, the Zionists interpreted the Balfour declaration as support for the establishment of their Jewish state in Palestine, and they believed that with Jewish immigration to Palestine that they would establish a Jewish majority within 10 years in Palestine, and they would then be able to establish their planned Jewish state. The Palestinians could hardly have welcomed the Zionist plan to make them a minority in their own country, and they were not even consulted about the plan or whether they would agree to the formation of a Jewish state in all or part of Palestine. The Balfour declaration was later incorporated into the British mandate over Palestine that was adopted by the League of Nations (the forerunner of the United Nations) in the early 1920s, and upgraded to a British commitment for a

Jewish homeland. There were riots in Jerusalem and Jaffa in 1920 and 1921, as Palestinians protested against plans for a Jewish home in Palestine.

The United Nations (the successor to the League of Nations) decided to partition Palestine into a Jewish part and an Arabic part in 1947, as after the Second World War there was international support for the creation of the Jewish state of Israel. This new state would enable survivors of the Nazi genocide of six million Jews in the Holocaust to live in their own country. The Palestinians did not accept the UN plan, and several Arab countries invaded but they were quickly defeated by Israel in 1948. Many Palestinian refugees (estimates of over 700,000) went to neighbouring countries during the violence of 1948, with some being expelled from their homes by violence and intimidation (e.g., the massacre of many inhabitants of the Palestinian village of Deir Yassin by Jewish paramilitary organisations such as the Irgun and Lehi) leading to fear and despair among the Palestinian community.[1]

Unfortunately many of the problems in the Middle East today are as a result of the way that the break-up of the former Ottoman Empire was handled by Britain and France after the first world war. Further, there cannot be lasting peace in the Middle East until the historical injustice suffered by the Palestinians is addressed in some reasonable way, and thus there is a need to find a fair and just agreement that would allow both sides to live with in peace and harmony together.

Israel has annexed more and more Palestinian territory since the Six-Day Arab–Israeli War of 1967, including large parts of the West Bank, East Jerusalem and the Golan Heights. The West Bank has an area of 5000 square km and a population of roughly 2.5 million. However, Israel has built roughly 150 settlements in the West Bank (including East Jerusalem) with a population of roughly 700,000). The expansion of the Israeli state onto the occupied Palestinian land after the 1967 war is illegal under international law, i.e., *all of these settlements in the West Bank, East Jerusalem and the Golan Heights are illegal under international law*. A two state solution looks impossible unless these settlements are handed over to Palestine as part of a peaceful resolution of the conflict.

Around 2 million Palestinians are crushed into roughly 400 square km in the Gaza strip, with Gaza city having a population of roughly half a million in an area of 45 square km. Noam Chomsky has done important work in the field of linguistics, and he is also an outspoken critic of American foreign policy and the state of Israel. He has stated that the Occupied territories are much worse off than South Africa (under the Apartheid regime), and that Israel just wants to get rid of the people, and that it has employed repressive policies in the Occupied Territories to make life unliveable for the Palestinians [2].

On the one hand the achievements of Israel since 1948 have been extraordinary, and it is an affluent and modern society with a well-educated population, and lots of hi-tech industries. However, unfortunately its success has been achieved at a

---

[1] The controversial Israeli historian, Ilan Pappé, describes this as the ethnic cleansing of Palestine [1].

massive cost, and there is a deep sense of injustice among the Palestinian people, and they view the actions of Israel since the formation of the state of Israel as deeply unjust and a violation of their rights. Bertrand Russell had sympathy to the plight of the Palestinians and he was critical of Israel.[2] He made several insightful remarks on Israel shortly before his death in 1970 [3]:

> The aggression committed by Israel must be condemned, not only because no state has the right to annexe foreign territory, but because every expansion is an experiment to discover how much more aggression the world will tolerate.

Social media have become an important communication channel for educated young people to discuss their aspirations for the future, as well as their grievances with society and the state. The effectiveness of Facebook as a tool for mobilising protests and social revolution is evident in the relatively short protests that culminated in the resignation of President Hosni Mubarak of Egypt in 2011. Egypt has a young population with roughly 60% of the population under the age of 30, and the country has faced many challenges since independence such as improving education and literacy for its young population, as well as finding jobs for its citizens. There are conflicts between modernity and tradition in many Islamic societies, as well as gender inequality issues, problems with the education systems, and a lack of political freedom with many autocratic political systems rather than democratic societies.

Facebook provided a platform for Egyptian youth to discuss issues such as unemployment, low wages, police brutality and corruption. Young Egyptians set up groups on Facebook to discuss specific issues, including a group to provide solidarity with striking workers. Further momentum for revolution followed the beating and killing of Khalid Mohammed Said, as photos of his disfigured body were posted over the Internet and went viral. An influential Facebook group called *"We are All Khalid Said"* was set up, and his killing provided a tangible focus for solidarity among young Egyptians.

Major protests broke out and lasted for eighteen days, and it led to hundreds of thousands of young Egyptians taking to the streets and gathering in Tahrir Square in Cairo. They demanded an end to police brutality as well as the end of the thirty-year reign of President Hosni Mubarak. The authorities reacted swiftly in closing the Internet in Egypt, but this act of censorship failed to stop the demonstrations and protests. Social media played an important role in mobilising protests, and in influencing the outcome of the revolution. President Mubarak resigned and he was later convicted of corruption, and President Morsi was inaugurated as the new

---

[2] Israel seems to view any criticism of the state of Israel as anti-Semitism (which is nonsense). Any solution to the Palestinian / Israeli conflict requires justice for the Palestinians, and it seems reasonable to an impartial observer that a just solution would be a Palestinian state based on the 1967 borders, and a return of the illegal Israeli settlements in the Occupied Territories. A just solution to the conflict would provide long-term security to the state of Israel, and an end to all the senseless violence.

president. The military removed Morsi from power in 2013, as his Muslim brotherhood government and autocratic style were deeply divisive leading to massive public discontent, and Abdel Fattah El-Sisi (a former military officer) replaced him.

6.2.2 Social Media Campaigns

Facebook is an effective tool for conducting various campaigns such as a social media marketing campaign to build awareness of a brand, or a mechanism where the customer can rate a particular product, or a way to improve customer loyalty by providing a forum where customers may discuss what they like or dislike about a product or company. Another words, a company may use Facebook as a tool so that its customers feel heard, and the company can show that it is listening to its customers, and responding to their feedback and improvement suggestions.

A social media campaign needs to be properly designed, and it needs clear goals on what it is trying to achieve as well as criteria for success. For example, the goals of a social media marketing campaign may be to communicate the company brands to a wider audience and to increase sales. The campaign may involve the participation of *influencers*, who are bloggers with a large following, to produce favourable content such as videos, images and articles on the product or company. The fact that the influencers have a strong following helps in building up a brand presence and awareness, and ensures that the message is spread to a wider audience. There is a need for care in the choice of the influencers to ensure that they are the right fit for the marketing campaign.

Facebook is an effective tool for politicians in their election campaigns, and one of its main advantages is that it enables the politician to interact directly with their constituents and the electorate, while bypassing the mainstream media. In a sense, this helps the politician to avoid the scrutiny that they are generally subject to in traditional media, where one of the important functions of traditional media is to hold politicians accountable to the public. Journalists, researchers and presenters fact check the claims of politicians to ensure the accuracy of their communication, and to ensure openness and transparency of political discourse. Facebook provides a forum for politicians to present and communicate their message and policies, as well as enabling them to show their human side as well as their personal interests.

6.2.3 Facebook and Privacy

Facebook has a comprehensive set of terms and conditions that users must accept to use the site, and these include invasive privacy practices that give Facebook the right to collect data on user activity on the site to improve the user experience and also for commercial purposes. A user must click acceptance of the associated terms and conditions to use the site, and most users just accept, as this is the only option that they have if they wish to join the Facebook community. Another words, although users create their own personal content that content is not private, and Facebook is continuously monitoring their behaviour on the site.

Further, many 3rd party apps interface with Facebook through Facebook's Open Graph platform, and all these apps have their own privacy settings and collect data on the users. The data gathered by Facebook may be used to build up profiles of users including what they like and dislike, and this data is potentially valuable to advertisers or those involved in marketing or political research. We will discuss the Cambridge Analytic scandal in the next section, where the personal information of millions of Facebook users was misused.

Facebook encourages users to disclose more and more information about themselves, and the user data may be visible and searchable. There is a danger that information that has been shared in the individual's network could be shared more widely by one of her "friends". This could result in posts that should remain restricted to a small audience being spread widely with potential negative effects on the individual. Another words, users need to be prudent in the content that they share in order to stay safe on the site.

Further, a Facebook user may be accessing the site through a smartphone, and so the location of the user is known at all times, and an analysis of the location data allows the user's home and work locations to be determined, as well as their day-to-day life patterns.

There have been several security and data breaches on Facebook where the personal data of Facebook users were exposed. For example, a data breach that occurred in April 2019 exposed the records of half a billion Facebook users on Amazon cloud servers, including the personal data of their friends, their names, passwords, and email addresses. The phone numbers of over 200 million Facebook users was exposed on an open platform in September 2019.

6.2.4 The Cambridge Analytica Affair

The Cambridge Analytics affair refers to a scandal that became public after the victory of Donald Trump in the 2016 US presidential election campaign. Cambridge Analytica acquired the private Facebook data of roughly 30–90 million users[3] in 2014, and used the data to create psychological profiles of tens of millions of American voters. The objective was to then sell these profiles to political campaigns so that they could be used for political advertising, and some political analysis assistance was provided to the Ted Cruz campaign and later to the Trump campaign. However, it is unclear whether this approach was a deciding factor in

---

[3] The New York Times estimated that the private data of 50 million Facebook users were taken, whereas Facebook's CTO (Mike Schroepfer) later revised this estimate to approximately 87 million (with 70 million from the US). Cambridge Analytica itself estimated that it had obtained the profiles of 30 million users.

the Trump victory, as there were several other factors that led to the defeat of Hilary Clinton.[4]

A third party app ('*This is your digital life*') was employed to collect the private data of Facebook users for Cambridge Analytica, and this app used Facebook's Open Graph platform to interface with Facebook. The app was an academic research study that used an informed consent process, where several hundred thousand Facebook users gave their consent to the collection of their personal information, and completed an online survey for which they received a small payment. However, the app was also able to obtain both the private data of the survey participants as well as the private data of all their friends through the Facebook Open Graph platform.

This led to Cambridge Analytic acquiring the private data of tens of millions of Facebook users, which contained sufficient detail to enable Cambridge Analytica to create psychological profiles of those that it had data on including their location. Cambridge Analytica began to create individual psychological profiles in 2016 when they were hired to support Ted Cruz's presidential campaign. The Facebook information was invaluable to political campaigns, as it allowed them to consider each person's psychological profile, which suggested what type of advertisement would be most effective in influencing a particular person in a particular location.

The Trump campaign used the Facebook data to create psychological profiles on users, and they used the data to determine the user's personality based on their Facebook activity. This information was then used for sending customised messages about Trump to different users, and the advertisements were segmented into different categories depending on whether the user was a Trump supporter or a potential swing voter. For example, supporters of Trumps received positive visuals of him and information on voting stations, whereas swing voters received negative information and visuals on Clinton and her policies, as well as positive stories about Trump and his policies (Fig. 6.3).

The scandal with Cambridge Analytic's misuse of Facebook data caused a major sensation when it became public, and Zuckerberg apologised to the media on behalf of Facebook in 2018, and pledged to make changes. The Federal Trade Commission imposed a $5 billion fine on Facebook for the data breach in 2019, and the US Securities and Exchange Commission (SEC) imposed a $100 million fine for misleading investors on the risk impacts of data misuse in 2019. Facebook announced that it would comply with the importan6t European GDPR privacy law (see Chap. 5) in 2019.

---

[4] There were several reasons for Hilary Clinton's defeat where she was leading in the opinion polls in the weeks before the election. The Clinton campaign was over confident, the campaign underestimated the electorate's desire for change from the Obama era, and James Comey's decision to reopen the FBI investigation into the Clinton private emails did not help. Clinton lost the white working class vote to Trump in the rust belt area in the US, and African American voters and Millennial voters were unenthusiastic for her. The Clinton campaign was seen as focusing on the professional class rather than the working class and their needs such as employment. She was seen as an elite and divisive candidate, and perhaps gender was also a factor in the election.

Fig. 6.3 Cambridge Analytica Scandel

6.2.5 Facebook Groups and Moderating Content

Individuals or businesses may set up Facebook groups, and these groups may be public or private. The posts of a private group are restricted to the members of the group, whereas the posts of public groups available to all. The use of groups offers a way for an organisation or business to build a community around its brand and to develop relationships with its customers, and while a Facebook page allows a business to present the profile of the company and posts it wishes to share, a Facebook group allows users to join as members and engage in interaction with the company and with other members. This allows the business a constructive way of interacting with its customers, and allows it to determine what its customers are saying about it, and what they like or dislike about the products and services that it provides.

An individual may set up a Facebook group in an area that she is interested in, and the various special interest groups will be of interest to different communities. There may be a need to manage the community content to ensure that all posts use appropriate language and respect the dignity of the members, and online tools may be employed to flag any language that may be inappropriate. There may be a need to develop a code of conduct for members of the group, and there be a need for a community facilitator to handle any issues that might arise.

That is, a role such as a community manager or administrator may be required to monitor the posts to ensure appropriate and ethical behaviour from the active community. Sometimes members may become highly emotive and engage in flame wars thereby upsetting other members (e.g., as in political debate). There could be *trolls*[5] inflaming hate, racism or other threatening behaviour affecting other members (especially in the debate of sensitive political topics such as immigration, gun laws, abortion rights, and LGBT rights).

There could be online harassment of members, where repeated insults or online bullying of an individual takes place, or threats of violence or even revenge porn. There may need to be discussions with violators of the terms and conditions of membership of the group, and in a worse case scenario there may be deletion of inappropriate posts and the removal of abusive members from the group.

---

[5] Trolling is where someone posts or comments online with the goal of upsetting other people.

6.3 The Tweet

Twitter/X is a social communication tool that allows people to broadcast short messages. It is often described as the *"SMS of the Internet"*, and Twitter/X is an online social media and micro blogging site that allows its users to send and receive short (initially 140-character) messages called *"tweets"*. The restriction to 140 characters was to allow Twitter to be used on non-smartphone mobile devices, but the character limit was doubled to 280 characters in 2017. Twitter has over 300 million active users, and it is one of the most visited websites in the world. Users may access Twitter through its website interface, a mobile device app or SMS.

Jack Dorsey (Fig. 6.4) and others founded the company in 2006. Dorsey introduced the idea of an individual using an SMS service to communicate with a small group while he was still an undergraduate student at New York University. The word *"twitter"* was the chosen name for this new service, and its definition as '*a short burst of information*' and '*chirps from birds*' was highly appropriate.

Twitter messages are often about friends telling one another about their day, what they are doing, where they are, whey they are thinking and doing, and any notable events that are happening around them. Twitter has transformed the world of media, politics and business. It is possible to include links to web pages and other media as a tweet. News such as natural disasters, sports results and so on are often reported first by Twitter. The site has impacted political communication in a major way, as it allows politicians and their followers to debate and exchange political opinions.

The effectiveness of Twitter/X as a platform for political communication was very evident in the 2016 presidential campaign of Donald Trump and throughout

Fig. 6.4 Jack Dorsey at the 2012 time 100 Gala

his presidency. The term *"Commander-in-Tweet"* was coined to refer to Trump's major use of Twitter throughout his presidency. Twitter enables celebrities to engage and stay in contact with their fans, and it provides a new way for businesses to advertise its brands to its target audience.

A Twitter user may select which other people that they wish to follow, and when you follow someone their tweets show up in a list known as your *Twitter stream.* Similarly, anyone that chooses to follow you will see your tweets in their stream.

A *hashtag* is an easy way to find all the tweets about a topic of interest, and it may be used even if you are not following the people who are tweeting. It also allows you to contribute to the topic that is of interest, and a hashtag consists of a short word or acronym preceded by the hash sign (#). Conferences, hot topics, and so on often have a hashtag.

A word or topic that is tagged at a greater rate than other hashtags is said to be a *trending topic*, and a trending topic is often the result of an event that prompts people to discuss the topic (e.g., an earthquake, the death of a celebrity, a political event). Trending may also arise from the deliberate action of certain groups (e.g., in the entertainment industry) to raise the profile of a musician or celebrity and to market their work.

Twitter has evolved to become an effective way to communicate the latest news, and its effectiveness as a communication tool for an organization increases as the number of its followers grows. An organization may determine what people are saying about it, as well as their positive or negative experience in interacting with it. This allows the organization to directly interact with its followers, which is a powerful way to engage with its audience and to make people feel heard. It allows the organization to respond to any negative feedback, and to deal with such feedback sensitively and appropriately.

The first version of Twitter was introduced in mid-2006, and it took the company some time to determine exactly what type of entity it was. There was nothing quite like it in existence, and initially it was considered a micro-blogging and social media site. Today, it is viewed as an information network rather than just a social media site.

Twitter has experienced rapid growth from 400,000 tweets posted per quarter in 2007, to 100 million per quarter in 2008, to 65 million tweets per day from 2010, to 140 million tweets per day in 2011 and to 500 million tweets per day since 2016. Twitter's usage spikes during important events such as major sporting events, natural disasters, the death of a celebrity, and so on. For such events, there may be over 100,000 tweets per second.

Twitter's main source of revenue is advertisements through *"promoted tweets"* that appear in a user's timeline (Twitter stream). The first promoted tweets appeared from late 2011, and the use of a tweet for advertisement was ingenious. It helped to make the advertisement feel like part of Twitter, and it meant that an advertisement could go anywhere that a tweet could go. Advertisers are only charged when the user follows the links or re-tweets the original advertisements.

Further, the use of tweets for advertisement meant that the transition to mobile was easy, and today over 80% of Twitter use is on mobile devices.

Twitter has recently embarked on a strategy that goes beyond these advertisements to sell products directly (including to people who don't use Twitter). Twitter also earns revenue from a data licensing arrangement where it sells its information to companies who use this information to analyse consumer trends. Twitter analyses what users tweet to understand their intent. Twitter was taken over by Elon Musk in 2022 and the company has undergone major restructuring. The company is currently known as Twitter/X or just X. For more detailed information on Twitter see [4].

6.4 LinkedIn

LinkedIn is a social media site for professional networking, and it allows professionals to create a profile of their qualifications and professional experience. The user's profile provides a short executive summary of the qualifications, knowledge and experience of the individual, as well as what sets him or her apart from others. It includes a selection of their professional experience and previous roles, and any distinguishing accomplishments, as well as recommendations from work colleagues and other professionals.

LinkedIn allows members to connect to other members of the site, and the user's connections form his or her professional network. These may include work colleagues and other professionals, and the site helps individuals in their interactions with other professionals, and allows them to post updates (e.g., new achievements, new roles, etc.) to enhance their professional reputation. LinkedIn includes a messaging service, where you can send messages to others who are both inside and outside of your network.

Employers may create profiles and become members of the site, and they may post job vacancies on the site. The site allows users to search and apply for new positions, and individuals may share their resume with potential employers and apply for relevant vacancies. Both employers and job seekers use LinkedIn widely, and there are over 900 million members of the site.

LinkedIn allows special interest groups to be formed to discuss special interest topics, and these groups may be public or private. LinkedIn's revenue comes mainly from selling access to information about its members to recruiters, and the company became a subsidiary of Microsoft in 2016. The basic version of LinkedIn is free, but LinkedIn Premium is a subscription version that offers additional features such as online classes and seminars.

6.5 Fake News and Disinformation

Fake news is the systematic spreading of false or misleading information in traditional print or online social media, with the intention of misleading or damaging another person or institution. It may negatively affect individuals, and could lead to violence or hate against minority or ethnic groups. The popularity of social media sites such as Facebook have contributed to the spread of fake news, and this new phenomenon poses threats to twenty-first century democracy. Fake news may be spread by individuals, organisations and hostile states, and it consists of news that has no basis in fact, but which is presented as being factually correct.

Fake news in the form of propaganda has been around for centuries, where such news is generally published for political reasons. Military leaders have often embellished their bravery and results in battle throughout history (e.g., the pharaoh Ramses II's description of the thirteenth century B.C. battle of Kadesh between the Egyptians and the Hittites paints a very positive (but factually inaccurate) account of the battle.

Following the invention of the printing press in the fifteenth century, news publications became popular, and over time fake news stories appeared in the print media. Fake news played an important role in propaganda during the first and second world wars, with radio broadcasts and printed material used to encourage and persuade the public at home as well as discouraging enemy troops. Today, modern society is highly dependent on accurate information in print, radio, television and on-line media. The effectiveness of fake news increases when the stories spread widely (as often occurs in social media), and where users interact with and rely on these stories rather than on traditional news media.

Fake news played an important role in the 2016 presidential election in the United States, which led to the election of Donald Trump. Most of the fake election news in the last three months of the campaign was anti-Clinton, but it is difficult to determine the extent to which this influenced the outcome of the election.

Trump and his supporters have used the term "fake news" to refer to the news from mainstream media that they accused of being opposed to him and his policies. The period in which Trump was in power was surreal, where the world as perceived by Trump did not correspond to the world of rational thinking human beings. The final denouement were the events on January 6th 2021 where Trump incited his supporters to march on Capitol Hill to claim back the country from the November 2020 "stolen election" victory of Biden. The extraordinary thing is that many of his supporters continued to believe that the election was stolen long after Biden was sworn in as president.

It is important when considering the accuracy of an article to consider the source of the news (e.g., is it written by a reputable news organisation such as the BBC or Reuters?), as well as considering the authenticity and reputation of its authors and the supporting sources. Fake news is a deeply disturbing trend that needs to be resolved if technology is to serve humanity, and if important democratic values are to be preserved in society. Modern technology has provided many benefits to modern society, but technology needs to serve humanity and not damage it. Some of the ways to check whether fake news might be present include:

- Consider the source of the news
- Check the author
- Check the date
- Check any supporting material
- Consult a fact checking site

Fake news is a dangerous trend in society, as false news can spread easily due to the speed and accessibility of modern technology. It allows individuals to be misled and adversely influenced, and individuals who rely on social media sites exclusively for their news are susceptible to being misinformed. Online social media sites such as Facebook and Twitter have a responsibility to develop appropriate solutions to address this dangerous problem, and following Trump's incitement of his followers to storm the capital in January 2021 they finally took appropriate action. Facebook and Twitter finally suspended Trump's account shortly before Biden's inauguration as a response to the excesses of Trump's misuse of social media.

There is a need for more rigorous terms and conditions on the use of social media, and social media sites have a professional and ethical responsibility to ensure that there is an appropriate balance between the right to free speech and the prohibition of disinformation, hate speech, racism, on line bullying, and so on. Social media sites need to take firm action when their terms and conditions are abused, and repeat offenders need to be blocked from their sites.

Social media is extremely useful but it creates new problems for society. It takes time for the legal system in a state to catch up with the problems caused with the introduction of a new technology, There is a need for the regulation of social media sites to deal with these problems, and also to deal with the fact that in a sense a social media site is a media organization, and so in a sense has additional responsibilities that need to be managed as such. That is, a social media site needs to take responsibility for whatever is published on its site, and to immediately respond and remove any inappropriate material (fake news, hate speech, etc.) from its site.

6.6 Data Analytics for Social Media

Data analytics provides a quantitative insight into human behaviour on a site, and is a way to understand users and how to communicate with them better. It enables the business to understand its audience better, and to create content that will be of interest to them. Data analytics consist of a collection of data that says something about the social media conversation, and it involves the collection, monitoring, analysis, summarisation, and graphs to visualise insight into the behaviour of users.

That is, *data analytics* involves learning to read a social media community through data, and the interpretations of the quantifiable data (or metrics) gives information on the activities, events, and conversations. This includes what users like when they are online, but other important information such as their opinions and emotions need to be gathered through *social listening*. Social listening involves monitoring keywords and mentions in social media conversations in the target audience and industry, to understand and analyse what the audience is saying about the business and allows the business to engage with its audience.

Social media companies use data analytics to gain an insight into customers, and elementary data such as the number of likes, the number of followers, the number of times a video is played on youtube, and so on are gathered to obtained a quantified understanding of a conversation. This data is valuable in judging the effectiveness of a social media campaign, where the focus is to determine how effective the campaign has been in meeting its goals. The goals may be to increase the number of users or to build a brand, and data analytics combined with social listening help in understanding how people are interacting, as well as how successful the interactions has been.

Facebook and Twitter maintain a comprehensive set of measurements for data analytics. Facebook maintains metrics such as the number of page views and the number of likes and reach of posts (i.e., the number of people who saw posts at least once). Twitter includes a dashboard view to summarise how successful tweet activity has been, as well as the interests and locations of the user's followers. Social listening considers user opinions, emotions, views, evaluations, and attitude, as well as a rich collection of human emotions.

The design of a social media campaign involves determining the objective of the campaign and designing the campaign to meet the requirements. The effectiveness of a campaign is judged by a combination of social media analytics and social listening, with the campaign refined appropriately to meet its goals and the cycle repeating. The key performance indicators (KPI) may include increased followers/subscribers or an increase in the content shared, and so on.

6.7 Ethics and Social Media

Facebook conducted a famous mood experiment on its users in 2012 to determine the extent to which their mood would be influenced by the content of their news feed. The experiment showed that happy stories generally led to positive postings, whereas sad stories led to negative postings, and so they were able to deduce that the mood of the news feed influenced the user's mood and postings. However, the experiment led to controversy, as users were unaware that they were participants in a psychological experiment, and the experiment had been conducted without their consent. The Facebook mood experiment involved 700,000 users and it showed a lack of courtesy and respect to its users, and it was also unethical since it was performed without the knowledge or consent of the affected users.

The principle of *informed consent* is fundamental in social media ethics, and individuals should be informed about an experiment, and have the right to agree to participate or decline their consent. Further, the individual should have the right to withdraw their consent at any time during the experiment. Once individuals have given their informed consent research data may be collected.

Social media sites collect data about the user's activities on the site, and this personal data may be processed and sold commercially to advertisers. Further, third party applications may interface with the social media site and gather personal data for commercial purposes. Users give consent to the social media platforms to collect and process their personal data, and their consent is given by clicking acceptance of the terms and conditions of use for the site. The consent cannot be said to be given voluntarily in the sense that if a user wishes to use the platform or third party application then she needs to click acceptance of the terms and conditions of use. Further, the consent cannot be said to be informed as the details of the data collection and processing are hidden in the small print of the terms and condition, which is generally a long unreadable document.

Those who create an online community (e.g., a community on Facebook) have a duty of care to the community that they have formed. This requires good community management to ensure the welfare of its members, and monitoring of activities to prevent abuse or online harassment of members of the community, and to ensure appropriate community behaviour at all times.

Ethical social media should consider the diversity of users and their needs, with appropriate accessibility interfaces provided for the visually impaired or those with a hearing impairment. Social media sites are designed to be addictive and to encourage users to disclose more and more information about themselves.

Social media sites have a professional and ethical responsibility to ensure that groups that advocate extremism or violence be banned from the site, and that the activities of such groups be reported to the relevant authorities.

There is a need for the regulation of social media sites as a social media site is essentially a media organization, and so it needs to take responsibility for whatever is published on its site, and to immediately respond and remove any inappropriate material (fake news, hate speech, etc.) from its site.

6.8 Review Questions

1. What is social media?
2. What are the advantages and disadvantages of social media?
3. Explain the role that social media played in the Arab spring.
4. What is a social media campaign?
5. Explain the significance of the Cambridge Analytica affair.
6. What privacy issues have arisen with social media?
7. What is the purpose of moderating content on social media?

8. What is Twitter?
9. What is LinkedIn?
10. Explain the significance of fake news.
11. What are data analytics?

6.9 Summary

Social media involves the use of computer technology for the creation and exchange of user-generated content, and it has led to major changes in communication between individuals, communities and organisations. Users may discuss and modify user created content, and social media plays a key role in connecting people and enabling communication.

Politicians use social media during election campaigns as a way to communicate their message and policies to a wider audience. Corporations use social media to understand what customers like and dislike about their products and services, with the goal of improving to serve them better. Social media enables citizens to play a role in shaping the world around them.

Facebook is a powerful communication tool that enables users to send and receive messages, and to keep in touch with friends and family around the world. It allows users to create and share information, and they may upload photos and videos, express opinions and ideas, exchange messages, and collaborate. A Facebook community is consists of a group of like-minded individuals that are interacting for a common goal. Twitter is a social communication tool that allows people to broadcast short messages called "tweets". It has over 300 million active users, and users may access Twitter through its website interface, a mobile device app or SMS.

Fake news is the systematic spreading of false or misleading information in traditional print or online social media, with the intention of misleading or damaging another person or institution. It may be spread by individuals, organisations and hostile states, and it consists of news that has no basis in fact, but which is presented as being factually correct. The popularity of social media sites such as Facebook have contributed to the spread of fake news, and this new phenomenon poses threats to twenty-first century democracy.

References

1. I. Pappé, *The Ethnic Cleansing of Palestine* (Oneworld Publications, 2006)
2. N. Chomsky, *Chomsky on Israel Apartheid, Celebrity Activists, BDS and the One-state Solution.* Middle East Monitor, 27th June 2022
3. *Bertrand Russell's Last Message on Palestine and Israel.* 31st Jan 1970. Read aloud in Cairo on Feb 3rd (1970)
4. M.W. Schaefer, *The Tao of Twitter. Changing your Life and Business 140 Characters at a Time*, 2nd edn (McGraw-Hill, 2014)

Ethics and AI

Abstract

This chapter discusses ethics and AI, and we discuss Weizenbaum's Eliza program and the challenge that AI poses to human dignity. We discuss the ethics of self-driving cars, and the need to encode self-driving vehicles with an appropriate moral compass to deal with situations where ethical decisions need to be made. We discuss ethical problems that arise with AI as well as ethical problems with expert systems.

Keywords

Human dignity • Self-driving cars • Expert systems • Surveillance • Privacy • Robotics • Intelligent machines

7.1 Introduction

The long-term[1] goal of Artificial Intelligence may be to create a thinking machine that is intelligent, has consciousness, has the ability to learn, has free will, and is ethical. Artificial Intelligence is a young field and John McCarthy and others coined the term in 1956. It is a multi-disciplinary field, and its branches include logic, philosophy; psychology; linguistics; machine vision; neural networks and expert systems.

Alan Turing devised the Turing Test as a way to test the intelligent behaviour of a machine, and Turing believed that machine intelligence was achievable. Searle's Chinese Room argument is a rebuttal of strong AI, and it aims to demonstrate that

---

[1] This long-term goal may be hundreds of years as there is unlikely to be an early breakthrough in machine intelligence as there are deep philosophical problems to be solved. However, a lot of progress has been made in simulating intelligence with work on machine translation and machine learning algorithms.

a machine will never have the same cognitive qualities as a human even if it passes the Turing Test.

There are deep philosophical problems in Artificial Intelligence, and some researchers (including Hubert Dreyfus and John Searle) believe that its goals are impossible or incoherent. Even if Artificial Intelligence is possible there are moral issues to consider such as the exploitation of artificial machines by humans, and whether it is ethical to do this. Weizembaum[2] has argued that Artificial Intelligence is unethical.

McCarthy argued that human-level intelligence could be achieved with a logic-based system that provides declarative knowledge, and this specifies what something is rather than how it is determined. McCarthy's approach was rejected by proceduralists such as Minsky, who argued that the representation of knowledge should be done in some domain that embeds the knowledge in terms of procedures. Procedural knowledge is the knowledge exercised in the performance of some task such as riding a bicycle. Cognitive psychology is concerned with cognition and some of its research areas include perception, memory, learning, thinking, and logic and problem solving. Linguistics is the scientific study of language and includes the study of syntax and semantics.

Artificial neural networks aim to simulate various properties of biological neural networks. They consist of many hundred or thousands of simple processing units that are wired together in a complex communication network. Each unit or node is a simplified model of a real neuron which fires if it receives a sufficiently strong input signal from the other nodes to which it is connected. The strength of these connections may be varied in order for the network to learn to perform different tasks corresponding to different patterns of node firing activity.

An expert system is a computer system that allows knowledge to be stored and intelligently retrieved. It is a program that is made up of a set of rules (or knowledge). The domain experts generally supply the rules about a specific class of problems. Expert Systems include a problem solving component that allows an analysis of the problem, as well as recommending an appropriate course of action to solve the problem.

There are several stories of attempts by man to create life from inanimate objects: for example, the creation of the monster in Mary Shelly's Frankenstein. The monster is created by an over ambitious scientist who is punished for his blasphemy of creation (in that creation is for God alone). The Czech play "Rossums Universal Robots" is a science fiction play by Capek, and contains the first reference to the term "*robot*". The play considers the exploitation of artificial workers in a factory, and the robots (or androids) are initially happy to serve humans, but become unhappy with their existence over a period of time. The fundamental question that the play is considering is whether the robots are being exploited, and if

---

[2] Weizenbaum was a psychologist who invented the ELIZA program, which simulated a psychologist in dialogue with a patient. He was initially an advocate of Artificial Intelligence but later became a critic.

7.1 Introduction

so, whether this is ethical, and what should the response of the robots be to their exploitation. It eventually leads to a revolt by the robots and the extermination of the human race.

The origin of the term "Artificial Intelligence" is in work done on the proposal for the Dartmouth[3] Summer Research Project on Artificial Intelligence. John McCarthy and others wrote this proposal in 1955, and the research project took place in the summer of 1956.

The success of early AI went to its practitioners' heads and they believed that they would soon develop machines that would emulate human intelligence. They convinced many of the funding agencies and the military to provide research grants, as they believed that real artificial intelligence would soon be achieved. They had some initial (limited) success with machine translation, pattern recognition and automated reasoning. However, it is now clear that AI is a long-term project despite recent progress in simulating human intelligence. Artificial Intelligence is a multi-disciplinary field and includes disciplines such as:

- Computing
- Logic and Philosophy
- Psychology
- Linguistics
- Neuroscience and Neural Networks
- Machine Vision
- Machine learning
- Robotics
- Expert Systems
- Autonomous Vehicles
- Machine Learning and Machine Translation
- Epistemology and Knowledge representation.

The English mathematician, Alan Turing, contributed to the debate concerning thinking machines, consciousness and intelligence in the early 1950s [1]. He devised the famous "Turing Test" to judge whether a machine was conscious and intelligent. Turing's paper was very influential as it raised the idea of the possibility of programming a computer to behave intelligently.

Shannon considered the problem of writing a chess program in the late 1940s, and he distinguished between a brute force strategy where the program could look at every combination of moves, and a strategy where knowledge of chess could be used to select and examine a subset of available moves. The ability of a program to play chess is a skill that is considered intelligent, even though the machine itself is not conscious that it is playing chess. Modern chess programs have advantages

---

[3] Darmouth is a private university located in Hanover, New Hampshire.

over humans in terms of computational speed in considering the immense combinations of moves. The IBM chess program "Deep Blue" defeated Kasparov in 1997.

The area of ethics has become important in AI and we discuss Weizenbaum's views on how AI impacts human dignity, and discuss ethical concerns in the field of expert systems. We then proceed to discussing the ethics of self-driving cars, and areas such as AI and privacy, AI and unemployment and AI and surveillance. There is more detailed information on the AI field in Chap. 22 of [2].

7.2 AI and Human Dignity

Weizenbaum created the famous Eliza program at MIT in 1966, and the program interacted with a user sitting at an electric typewriter, in English, in the manner of a Rogerian psychotherapist.[4]

The name "*Eliza*"[5] was chosen because it could be taught to speak increasingly well, as in Shaw's play "Pygmalion", and its later adaptation in the 1964 film "My Fair Lady" starring Rex Harrison and Audrey Hepburn. Eliza was one of the earliest natural language processing programs, and Weizenbaum gave an example of a conversation between Eliza and a young woman in [3]. Many users believed that Eliza had real understanding, and they began to unburden themselves in long computer sessions.

The program operated by decomposing the user's input into its constituent parts of speech, and then typing them back at the user in a manner to continue with the dialogue. The program provided very convincing human-like interaction, as may be seen in the sample dialogue below (Fig. 7.1).

User/Patient:	Men are all alike
Eliza	*In what way?*
User/Patient:	They are always bugging us about something or other
Eliza	*Can you think of a specific example?*
User/Patient:	Well, my boyfriend made me come here
Eliza	*Your boyfriend made you come here?*
User/Patient	He says I'm depressed much of the time
Eliza	*I am sorry to hear that you are depressed*
User/Patient	It's true, I am unhappy
Eliza	*Do you think that coming here will help you not to be unhappy.?*
User/Patient	I need some help, that much seems certain
Eliza	*What would it mean to you if you got some help?*

(continued)

---

[4] Rogerian psychotherapy (person-centred therapy) was developed by Carl Rodgers in the 1940s.
[5] Eliza Doolittle is a working-class character in Shaw's play "Pygmalion". She is taught to speak with an upper-class English accent.

(continued)

User/Patient	Perhaps, I could learn to get along with my mother
Eliza	*Tell me more about your family*

Weizenbaum was shocked to discover that so many users were convinced that the program had real understanding, and users spent hours sharing their personal problems with the program. It led Weizenbaum to think about the ethics and implications of the artificial intelligence field given the ability of a relatively trivial program to deceive a naïve user to reveal personal information. He became a leading critic of the AI field, and an advocate for professional and social responsibility in computer science.

His views on the ethics of AI are discussed in his book "Computer Power and Human Reason" [4]. He displays ambivalence towards computer technology, and he argues that AI is a threat to human dignity, and that AI should not replace humans in positions that require respect and care. He states that machines lack empathy, and that if they replace humans in positions such as police officers or judges, that this would lead to alienation and a devaluation of the human condition.

His Eliza program demonstrated the threat that AI potentially poses to privacy. It is conceivable that an AI program may be developed in the future that is capable of understanding speech and natural language in such a way that it could theoretically eavesdrop on every phone conversation and email, and gather private information on what is said, and who is saying it.[6] Such a program could be used by a state to suppress dissent, and to eliminate those who pose a threat.

7.3 Ethics of Self-driving Cars

A driverless car (autonomous vehicle) is a vehicle that can sense its environment and navigate its way without human intervention. It uses techniques such as AI, GPS, radar and computer vision to detect its environment, and it has advanced control systems to determine an appropriate navigation path to its destination. Its navigation needs to be sophisticated to enable it to avoid obstacles, and to observe road signage and traffic lights during the journey, as well as dealing with diverse weather/light conditions (Fig. 7.2).

The control system includes sensing and navigation systems, and the analysis of the sensory data must be able to distinguish between different vehicles on the road. The control system must make the correct decisions from the analysis of the images, and this is especially important when dealing with unexpected situations.

Several technology companies such as Google, Apple, Uber and Amazon are working on driverless cars, and autonomous vehicles may potentially lead to a significant reduction in road accidents and fatalities. They offer greater mobility

---

[6] One wonders what information the virtual assistant, Amazon Alexa, gathers every day and how that information is used.

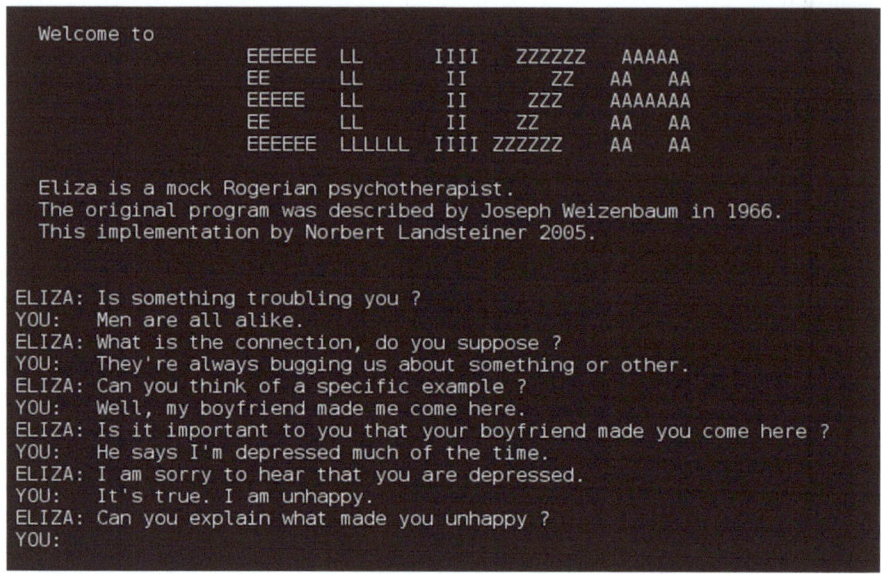

Fig. 7.1 Eliza program

Fig. 7.2 Waymo self-driving Car in California, 2017

for people who cannot operate a vehicle, but there are many challenges such as safety and security issues that need to be solved, as well as the development of an appropriate regulatory framework, before the public will have sufficient confidence in their use (Table 7.1).

Driverless cars will need to be encoded with a moral compass to deal with situations where ethical decisions needs to be made. For example, suppose a self-driving vehicle is travelling on a road and two children roll off a grassy bank on to the road. Further, there is no time for the vehicle to brake and so what should the vehicle do where there is a choice of protecting the pedestrian or the passenger? For example, if the vehicle swerves to the left to avoid the children it will hit

7.3 Ethics of Self-driving Cars

Table 7.1 Challenges with driverless vehicles

Area	Description
Sensing the Surroundings	A motorway looks totally different on a clear day than on a foggy day or at dusk. Driverless cars must be able to detect road features in all conditions, and the sensors need to be reliable
Unexpected Encounters	Driverless cars struggle with unexpected situations (e.g., traffic police waving vehicles through a red light), as rule based programming is unlikely to cover every scenario
Human Vehicle Interaction	Most self-driving cars will be semi-autonomous for the foreseeable future, and determining the responsibilities of human and machine and when one or the other should be in control is a challenge
Ethical	Should the car prioritise the protection of the pedestrian or the passenger? Moral judgments may be required
Security/ Hacking	Conventional vehicles have vulnerabilities that may be exploited by hackers (e.g., the braking and steering system of a vehicle was hacked through its entertainment system in 2015). Self-driving cars have more vulnerabilities and are at a greater risk of a malicious attack
Legal Framework/ Liability	Self-driving vehicles will be subject to strict safety regulations, and appropriate legislation needs to be developed

an oncoming motorbike. *Which decision should the car make and how should it make such a decision?* Further, who should be held accountable when incorrect or unethical decisions are made?

This is a variant of the *trolley problem*, which is a famous thought experiment in ethics. A train is rushing down a track out of control as its brakes have failed. Disaster lies ahead as five people are tied to the track and will perish in the absence of action. There is sufficient time to flick the points and divert the train down another track, where there is one man tied to the track. Is it ethical to divert the train to do this? Most people would be inclined to take the view that this is the best (least worst) outcome (Fig. 7.3).

There is a controversial variant of the problem where the train is rushing towards five people and you are standing on top of a footbridge overlooking the track next to a man with a very bulky rucksack. The only way to save the five people is to push the man to his doom, as his rucksack will block the train and

Fig. 7.3 Trolley problem

save the five. Is it ethical to deliberately kill or sacrifice another human being to save five others? Most people would say no to this deliberate killing, but it would be valid in the utilitarian school of ethics, which seeks to maximize happiness in the world.

Even though the trolley problem is a thought experiment, it is conceivable that a driverless car will face situations where a moral choice must be made (e.g., who to harm or injure such as pedestrians, passengers or driver). Clearly, this raises the importance of the type of ethics that are programmed into the car, and who is to decide on what are the appropriate ethics? How can you test that the vehicle has the desired ethics to operate safely in all situations that require an ethical judgment? How would an autonomous vehicle be held accountable for errors in decision-making, which lead to injury or loss of life?

A self-driving car must be able to deal with unpredictable human driving or behaviour, and the question is who should be held accountable when the self-driving vehicle makes incorrect decisions that lead to a loss of life? Should the designers of the software for the vehicle be held liable? Should the regulators be held responsible for allowing an unproven and potentially unsafe technology on a public road thereby endangering the public? Should the precautionary principle be strictly applied until it has been demonstrated that autonomous vehicles pose no safety risk to the public?

Self-driving vehicles are being tested worldwide, and there have been several fatalities. These include an Uber autonomous vehicle fatality in 2018 in Arizona, where the autonomous car killed a pedestrian. The precautionary principle was discussed in Chap. 3, and it argues that there is a burden of proof on the manufacturer to show that a technology is safe, and that the technology should not be used until it has been proved to be safe. Uber suspended testing of autonomous vehicles after the accident.

As more and more sophisticated machines and robots are created, it is, of course, essential that intelligent machines behave ethically, and have a moral compass to distinguish right from wrong. It remains an open question as to how to teach a robot right from wrong, and in view of the recent progress that has been made in AI, the time is approaching where machines will routinely make ethical decisions.

Teaching ethics may involve programming in certain principles, and then the machine learns from various scenarios on how to apply the principles to new situations. There is a need for care with machine learning as the machine may learn the wrong lessons, or since its learning evolves over time it may not be possible to predict its behaviour in the future. Further questions arise as to who is to be held accountable in the event of a machine making incorrect or unethical decisions. For further information on the feasibility of teaching ethics to machines see the interesting BBC article *"Can we teach robots ethics?"* [5].

7.4 Expert Systems and Ethics

An expert system is a computer system that attempts to copy the reasoning process of human experts, and it contains the domain knowledge of one or more human experts in a narrow specialised domain. It consists of a set of rules (or knowledge) supplied by the domain experts about a specific class of problems, and it allows knowledge to be stored and intelligently retrieved. The effectiveness of the expert system is largely dependent on the accuracy of the rules provided, as incorrect inferences will be drawn with incorrect rules. Several commercial expert systems have been developed since the 1960s.

Expert Systems have been a success story in the AI field, and they have been applied to the medical field, equipment repair, and investment analysis. They employ a logical reasoning capability to draw conclusions from known facts, as well as recommending an appropriate course of action to the user. An expert system consists of a knowledge base; an inference engine; an explanatory facility; a user interface; and a database (Table 7.2).

Human knowledge of a specialty is of two types: namely public knowledge and private knowledge. The former includes the facts and theories documented in textbooks and publications, whereas the latter refers to knowledge that the expert possesses that has not found its way into the public domain. The latter often consists of rules of thumb or heuristics that allow the expert to make an educated guess where required, as well as allowing the expert to deal effectively with incomplete or erroneous data. It is essential that the expert system encode both public and private knowledge to enable it to draw valid inferences (Fig. 7.4).

The inference engine is made up of many inference rules that are used by the engine to draw conclusions. Rules may be added or deleted without affecting other rules, and this reflects the normal updating of human knowledge. Out of date facts may be deleted, as they are no longer used in reasoning, while new knowledge may be added and applied in reasoning. The inference rules use reasoning that is closer to human reasoning, and the two main types of reasoning are backward chaining and forward chaining. Forward chaining starts with the data available, and uses the inference rules to draw intermediate conclusions until a desired goal is reached. Backward chaining starts with a set of goals and works backwards to determine if one of the goals can be met with the data that is available.

Table 7.2 Expert systems

Component	Description
Knowledge base	The knowledge base is represented as a set of rules of the form (IF condition THEN action)
Inference engine	Carries out reasoning by which Expert System reaches conclusion
Explanatory facility	Explains how a particular conclusion was reached
User interface	Interface between user and expert system
Database/memory	Set of facts used to match against IF conditions in knowledge base

Fig. 7.4 Expert (Albert Einstein)

The expert system makes its expertise available to decision makers who need answers quickly. This is extremely useful as often there is a shortage of experts, and the availability of an expert computer with in-depth knowledge of specific subjects is therefore very attractive. Expert systems may also assist managers with long-term planning. There are many small expert systems available that are effective in a narrow domain.

Several expert systems (e.g., Dendral, Mycin, and Colossus) have been developed. *Dendral* (Dendritic Algorithm) was developed at Stanford University in the mid-1960s, and its objectives were to assist the organic chemist with the problem of identifying unknown organic compounds and molecules by computerized spectrometry. This involved the analysis of information from mass spectrometry graphs and knowledge of chemistry. Dendral automated the decision-making and problem-solving process used by organic chemists to identify complex unknown organic molecules. It was written in LISP and it showed how an expert system could employ rules, heuristics and judgment to guide scientists in their work.

Mycin was developed at Stanford University in the 1970s. It was written in LISP and was designed to diagnose infectious blood diseases, and to recommend appropriate antibiotics and dosage amounts corresponding to the patient's body weight. It had a knowledge base of approximately 500 rules and a fairly simple inference engine. Its approach was to query the physician running the program with a long list of yes/no questions. Its output consisted of various possible bacteria that could correspond to the blood disease, along with an associated probability that indicated the confidence in the diagnosis. It also included the rationale for the diagnosis, and a course of treatment appropriate to the diagnosis.

Mycin had a correct diagnosis rate of 65%. This was better than the diagnosis of most physicians who did not specialise in bacterial infections. However, its diagnosis rate was less than that of experts in bacterial infections who had a success rate of 80%. Mycin was never actually used in practice due to legal and ethical reasons on the use of expert systems in medicine. For example, if the machine makes the wrong diagnosis who is to be held responsible?

Colossus is an expert system that is used by several Australian insurance companies to help insurance adjusters assess personal injury claims. It helps to improve consistency, objectivity and fairness in the claims process. It guides the adjuster through an evaluation of medical treatment options, the degree of pain and suffering of the claimant, and the extent that there is permanent impairment to the claimant, as well as the impact of the injury on the claimant's lifestyle. It was developed in Australia in the late 1980s, and acquired by the Computer Sciences Corporation (CSC) in the mid-1990s.

It is important to consider the ethics of computerised expert systems, and an important question to consider is whether an expert system should be allowed to enforce a decision without human intervention? There are potentially serious consequences to incorrect decisions made by a computerised system, and who should be held responsible when poor decisions are made, or where the expert system causes harm?

Another words, who should be held accountable if the expert system issues incorrect decisions/recommendations? Should the software developers or the team that designed and developed the system be held accountable? Was expert knowledge properly encoded into the system? Was the system properly designed and tested? Was it piloted prior to general release? Were detailed operating procedures defined and implemented for the circumstances where the expert system may be used without human intervention?

The track record of the expert system may be used to build up confidence in the expertise and trustworthiness of a particular expert system, and it may be that some expert systems should never be used without human oversight. Further, as human knowledge is continuously evolving as new facts and inventions take place, an expert system needs to be upgraded on a regular basis to ensure that its analysis is up to date. Further, the users of an expert system should be informed as to how trustworthy or reliable the system actually is, so that they may determine the extent that they should rely on the expert system, or whether it would be more appropriate to seek the advice of a human expert.

7.5 AI and Unemployment

The industrial revolution in Britain was a major turning point in human development. It led to the transformation of traditional agrarian societies that used hand-based production methods to create products, to a new industrial world of technology driven by steam engines and the development of factories that mass-produced products. The industrial revolution transformed the way in which work was done, and it led to the organisation of labour into workers and supervisors / managers, and an increasing specialisation of the roles performed within manufacturing companies.

Fig. 7.5 Unemployed in Chicago in 1930s

Advanced information technology has led to the displacement of workers in many traditional industries, and today there is increasing mechanisation of manufacturing companies to improve productivity and save costs. The use of automation and robotics has accelerated this trend, and it has resulted in workers being displaced and needing to reinvent themselves for other categories of employment. However, the displacement of employees due to the introduction of new technologies often leads to the creation of totally new categories of employment. Globalisation with massive outsourcing of manufacturing and services has accelerated the displacement of employees, and often it is low skilled non-college workers that are displaced, leading to lower salaries for this cohort of workers, and lower living standards.

AI technology is likely to displace workers in a similar way to the introduction of earlier technologies, and so it is likely to lead to increases in unemployment in the short term. However, the long-term expectation is that higher-value roles will emerge, and that with appropriate education and training that workers will be free to perform these higher-values roles. There is also a role for human/AI collaboration as in augmentation intelligence, where AI is employed to assist humans in performing their roles more effectively. Clearly, humans would be much happier in working with machines to achieve enhanced performance rather than being replaced by them (Fig. 7.5).

7.6 AI and Surveillance

AI and facial recognition technology has facilitated the recognition of individuals at mass gatherings such as demonstrations, riots and public places. Further, GPS technology allows the locations visited by a person to be tracked, and so AI technology may become in a sense become an enabler of the Big Brother character in Orwell's totalitarian world of *Nineteen Eighty Four* (Fig. 7.6).

That is, a government could potentially use facial recognition technology for authoritarian control, and surveillance combined with facial recognition may be oppressive for society. This could lead to a world with technology monitoring the activities of citizens, learning to recognise patterns, drawing inferences, and taking action to prevent any civil disobedience.

There is a rich database of digital photographs available on social media websites; driver licenses registries, and so on. Faces are central to identity and faces are hard to hide, and facial recognition is a biometric technology that analyses visual data from social media and other sources. They essentially reduce each face to a mathematical equation using factors such as the distance between the individual's eyes, the width of the nose, and so on, and the patterns in the visual data are compared to patterns in existing facial recognition databases to confirm identity. Facial recognition is a potentially dangerous technology that may challenge civil liberties, and it could severely impact marginalized groups in society.

Surveillance is the observation or monitoring of a person, and includes visual observation as well as all other behaviours and actions. In the past, it was achieved through the use of traditional technologies such as surveillance cameras and wiretaps, but today this has been expanded to Internet and GPS surveillance. It is uncomfortable to have a stranger stare at you for an extended period of time, and

Fig. 7.6 Big Brother is watching you

this, in a sense, is what constant surveillance is, except that the individual may be unaware that he/she is being watched.

The usual argument made to defend surveillance is that law-abiding citizens have nothing to fear from or surveillance, and most citizens will not be targeted apart from those who are engaged in criminal activity. And so, surveillance in a sense has the potential to protect lives with minimal invasions of privacy. However, there is a danger of a slippery slope with surveillance, and limits need to be set on the acceptable bounds for its use.

Electronic surveillance is often used to fight serious crime, and to prevent terrorism as well as protecting state security. By nature, it is a secret and intrusive method for information collection, and it is performed without the knowledge of the surveillance target. The security services need to possess exceptional powers to carry out surveillance, and these could potentially infringe on fundamental human rights, especially the right to privacy. Therefore, since electronic surveillance may potentially harm the individual there is a need for safeguards on electronic surveillance such as comprehensive legislation, controls and oversight.

7.6.1 Biometric Technology

The term "Biometrics" consists of two parts namely "*bio*" as in biology and representing the study of living organisms, and "*metrics*" for the measurement of data. Biometrics is the measurement and statistical analysis of the unique physical and behavioural characteristics of humans, and the technology is generally used to identify individuals who are under surveillance as well as for access control purposes.

Biometrics is a way to ensure that the person is real and who they claim to be by verifying a real-world trait of the person. Biometric passports include an embedded microchip containing the user's personal information including a photograph, and this personal information is verified at passport control or at an e-password gate. Smartphone fingerprint unlocking has become a popular way to provide security on mobile phone devices, and it helps to ensure that only the owner of the smartphone may access it. There are several well-known techniques used by biometric technology including:

- Fingerprint recognition
- Face recognition
- Iris recognition
- Voice recognition
- Palm or finger recognition
- DNA-based recognition

Biometric technology works by capturing the biological input that the user provides, and the biometric software then measures the captured data to create a template or "lock" representing the data that will be used for future comparisons.

This template is then stored on the internal hardware or the cloud, and new inputs are then measured by biometric sensors and compared to the lock to confirm the person's identity and unlock the device. The biometric template (or "lock") is a code generated from the "lock" image, rather the scanned image and this adds a level of security protection, as if a hacker succeeds in gaining access to the biometric database then they would have access to the codes only but not the underlying biological data.

More and more business and government applications are using biometric authentication, and so they are collecting and storing massive amounts of personal biometric data. Therefore, there is a need to protect sensitive biometric data with increased security against hackers, and to consider the legal and ethical implications of the technology. Biometric data is permanent and cannot be concealed (eyes, face and fingerprints are unique to every individual throughout their life), and many technology companies have gathered the biometric data of millions of people in their authentications systems. This data will be there in perpetuity, and so it needs to be protected to ensure that the data subject is not harmed from its misuse.

The European GDPR was discussed in Chap. 5 and it places rules on the way that biometric data may be collected and used, and it emphasises the need for caution in using this sensitive data. The right to privacy is enshrined in Article 12 of the Universal Declaration of Human Rights, Article 8 of the European Convention of Human Rights, and in Article 7 of the European Charter of Fundamental Rights.

There are ethical questions on the collaboration and sharing of biometric data as in public–private collaboration. That is, general data protection principles apply such as what is the purpose of the data and whom should it be shared with? There is a need for regulation as well as independent oversight and governance of *live facial recognition software* (LFR), as well as maintaining an appropriate balance between the respect for the privacy of individuals and the benefits of the technology.

LFR compares a live camera feed (or multiple feeds) of faces against a particular face to locate a person of interest and to generate an alert when found. Biometric data should be kept secure at all times and protected from misuse by unauthorised individuals.

7.7 AI Algorithms and Discrimination

We discussed Weizenbaum's Eliza program from the 1960s earlier in the chapter, and noted the threat that AI poses to privacy. Future AI programs may be capable of eavesdropping on verbal and written electronic communication, and gathering private information on what is said, and who is saying it.

AI may use personal information in ways that may intrude upon privacy, such as in the way that it may analyse personal information with machine learning algorithms, search algorithms, recommendation engine algorithms, and algorithmic decision-making.

It is important to protect individuals from harmful use of their personal information in AI, and it is essential to ensure that AI algorithms are fair and are not biased. For example, the purpose of the *predictive policing algorithms* was to use machine-learning algorithms to identify people and locations at increased risk of crime, and they included methods for predicting crimes, offenders and victims. These algorithms used data on the times, locations and nature of past crimes to determine patterns, and to recommend where and when police should patrol or maintain a presence in order to make the most effective use of limited police resources. However, predictive policing algorithms have been criticised by civil rights groups due to their tendencies to proliferate racial profiling, and leading to the racial discrimination of minorities.

Amazon developed its *AI machine learning hiring algorithm* to mechanise its search for talent, but it later abandoned its efforts at mechanising recruitment due to biases in the algorithm. The computer algorithms were trained to vet applicants based on patterns observed in CVs submitted to the company over a ten-year period. However, as the majority of the hired staff was male the algorithm taught itself that male candidates were preferable to female, and it penalised CVs that included the word "women" as well as graduates of all female colleges. Amazon amended the AI program, but as there was no guarantee that the machine would not devise other ways of sorting the CVs that would not prove discriminatory to candidates it decided to terminate the project in 2015.

What should happen if a company or organisation uses an *algorithm that is biased*, and where the use of the personal data has an adverse impact on an individual? There is the potential for a biased algorithm to produce unlawful or undesired discrimination, and so the question is who should be held responsible when such discrimination happens? Is it the responsibility of the company that designed and implemented the algorithm? Is it the responsibility of the company using the algorithm? What role do legislators have in ensuring that the right laws have been enacted to prevent unlawful discrimination? Is there a need to regulate the way in which companies process the private data that they collect, as well as how they use it and whom they share it with? What role should civil liberties groups play in monitoring these developments?

It seems reasonable to argue that algorithmic decision-making based on personal information should be subject to existing civil rights and protections, and that there is a need for transparency and accountability. There may need to be an independent verification of the fairness of algorithms employed for critical decision-making.

7.8   AI and Disinformation

AI has facilitated the growth of fake content and *deep fake videos* involve carefully substituting one person for another in a video using artificial intelligence. It is becoming more difficult to spot and to distinguish the fake from the real person, and while they are often used for humour such as in caricaturing another person, they do potentially pose a threat to society. Deepfake technology involves taking a person's face and marking points around the face including chin, corners of mouth, corners of eyes, and then mapping those points to another person thereby swapping out one person for another.

There are tools available to detect deep fakes from genuine people based on learning patterns and unique characteristics of the individual, but to preserve trust in true footage one needs in a sense to authenticate the camera that took the footage. There are serious concerns of the dangers posed to society by deep fakes, as they could lead to fake news or disinformation, where a politician could appear to say something that he or she did not really say, or even incite followers to riot or even worse. However, to date the main application of deep fake technology has been to comedy and political satire, and social media sites such as Facebook are active in removing deep fake videos.

Political bots[7] and disinformation played a role in the 2016 Brexit referendum and in the 2016 presidential election in the United States. An Internet bot is a software application that runs automated tasks over the Internet, and it generally performs a simple and repetitive task much faster than a human. Although the software bots used in the 2016 election were quite simple and not AI applications, this may change in the future with more sophisticated applications. Many Twitter bots (believed to have been set up and carried out by a hostile foreign state) spread out false or manipulated stories during the 2016 campaign, using accounts that appeared to be from swing states in the Midwest. Often, people are inclined to believe a piece of information when it is coming from someone who appears to be local and looks similar to themselves, and this misinformation campaign was effective, and may have played a role in the victory of Trump in the 2016 presidential election in the United States.

7.9 AI and Autonomous Weapon Systems

Autonomous Weapon Systems (AWS) are a type of military system that can independently search for and engage military targets based on programmed constraints and descriptions. That is, such a system may select and engage military or civilian targets without the intervention of a human operator, and these systems may operate in the air, on land, under water, and in space. Such systems are capable of learning and adaptable to changes to the environment in which they are operating, and are capable of making final decisions on their own. Autonomous weapon systems include autonomous offensive systems and automatic defensive systems.

The term "autonomous" refers to the machine's ability to operate without human involvement, although in practice lethal autonomous weapon systems

---

[7] A chatbot is a computer program that simulates human conversation, and allows humans to interact with digital devices as if they were communicating with a real person. An Internet bot is a software application that runs automated scripts over the Internet.

(LAWS) are restricted in the sense that humans are required (with some exceptions for certain defensive systems) to give the final order to attack. The main reason for humans not being involved in the authorisation of an attack is where there is a need for a rapid or immediate response, such as the case where an imminent hostile attack has been identified that requires an immediate response in order to protect the lives of military personnel of a naval ship or submarine from an incoming missile attack.

Early examples of autonomous systems include landmines and naval mines that date from the seventeenth and eighteenth century respectively. Modern autonomous systems include active protection systems that are used to defend ships, where these systems can automatically detect and attack incoming missiles or other hostile attacks according to criteria set by the human operator. The use of drone technology may lead to a futuristic air defence system that can autonomously search, identify and locate enemies as well as defending itself against enemy aircraft and engaging in military action under authorisation from mission command.

Autonomous weapon systems continue to evolve and the modern battlefield may have very few people fighting, with warfare instead conducted by autonomous weapon systems. This means that the modern battlefield may consist of autonomous killer robots with lethal weapons attacking infrastructure and military personnel. In fact, drones are regularly employed in conflict zones around the world, and often with a devastating effect.

Therefore, it is important to consider the appropriate level of human control of autonomous weapon systems, and the Human Rights Watch Report outlines three possible levels of human control including:

- Fully autonomous systems (no human involvement)
- Monitored autonomous systems (A human may abort attack at any time)
- Initiated autonomous systems (A human initiates the attack)

There are legal and ethical problems with autonomous weapon systems including the problem of distinguishing between combatants and non-combatants. This has led some scholars to argue that such systems are inhumane and unethical, and some legal scholars have argued that these weapon systems are a violation of international humanitarian law.

It is essential that autonomous weapon systems that are capable of killing and that operate without human intervention be certified as compliant with international safety standards, and that autonomous weapon systems that are not compliant be banned. A group of over 1000 experts in the Artificial Intelligence field signed a letter in 2015 warning of the dangers of an Artificial Intelligence arms race, and calling for a ban on autonomous weapon systems.

7.10 Robots and Ethics

Asimov wrote several stories about robots in the 1940s including the story of a robotherapist. He predicted the rise of a major robot industry, and he introduced a set of rules (or laws) for good robot behaviour (*Laws of Robotics*). Asimov later added a fourth law (Table 7.3).

Robots have been very effective at doing clearly defined repetitive tasks, and there are many sophisticated robots in the workplace today. These are mainly industrial manipulators that are essentially computer controlled "arms and hands". They can improve the quality of life for workers as they can free human workers from performing dangerous or repetitive jobs. They provide consistently high-quality products and can work tirelessly 24 h a day. This helps to reduce the costs of manufactured goods thereby benefiting consumers.

An intelligent robot is a machine that can extract information from its environment, and use knowledge about the physical world to move safely in a meaningful and purposeful manner. The robot senses its environment and acts in a rational manner to achieve the defined goals. Robots require good engineering and science to be effective, including sensors, effectors/actuators, a locomotion system, an on-board computer system and various controllers.

The area of computer vision has made significant progress from the development of the Stanford Cart in the mid-1970s, and the Carnegie Mellon Rover and its successors from the early 1980s. The Stanford Cart was a simple buggy with a video camera, and Han's Moravev's version in the late 1970s could navigate slowly around a room with obstacles (without human intervention) in a controlled environment. The Carnegie Mellon Rover and its successors have been very successful, with the CMU Navlab 5 vehicle travelling from Pittsburg to Los Angeles in 1995 with an autonomous driving percentage of 98.2%.

Robots have been applied to many areas including assembly and manufacturing, industrial manipulation and materials handling, hazardous environments, space and underwater exploration in unmanned vehicles, education, remote environments, medical science, virtual reality and the entertainment industry. The advantages of robots include (Table 7.4).

Table 7.3 Asimov's laws of robotics

Law	Description
Law Zero (Humanity)	A robot may not injure humanity, or, through inaction, allow humanity to come to harm
Law One (Human Safety)	A robot may not injure a human being, or, through inaction, allow a human being to come to harm, unless this would violate a higher order law
Law Two (Slave)	A robot must obey orders given to it by human beings, except where such orders would conflict with a higher order law
Law Three (Survival)	A robot must protect its own existence as long as such protection does not conflict with a higher order law

Table 7.4 Advantages of robots

No	Advantage
1	Robots can do repetitive work $24 \times 7 \times 365$
2	Robots can do tasks that are too dangerous for humans
3	Robots can operate machinery to a much higher level of precision than humans
4	A robot may be able to perform tasks that are impossible for humans

Table 7.5 Disadvantages of robots

No	Disadvantage
1	Robots can make incorrect decisions in emergencies
2	Robots have limited movement and vision
3	Robots are costly and require programming and training
4	Robots may replace humans in the workplace leading to unemployment

The disadvantages of robots include (Table 7.5).

However, as the sophistication and intelligence of robots improve and as they begin to make decisions that impact humans, there is a need to ensure that robots behave properly and have an appropriate moral compass to distinguish right from wrong. The 1968 film "2001: A Space Odyssey" describes how the computer "HAL" that controls the systems of the spacecraft malfunctions, and attempts to kill the astronauts who have decided to shut it down.

7.11 Super-Intelligent Machines

A super-intelligent machine is a machine that possesses intelligence far exceeding the cognitive performance of the brightest of humans in virtually all domains of interest. In the modern world, sophisticated machines routinely outperform humans in specialised domains (this includes programs such as chess). This is especially the case where the speed of calculation is important, as computer hardware is much faster than the human brain in performing calculations. However, such programs fail to outperform humans in other activities, and so cannot be considered to be super-intelligent agents.

Several computer scientists (including David Chalmers) have argued that artificial super-intelligence is feasible, and Chalmers has argued that artificial general intelligence is a likely path to superhuman intelligence. His belief is that the first step is AI achieving the equivalence of human intelligence, and then it being extended to surpass human intelligence, and finally being further amplified to completely outperform humans across arbitrary activities [6].

He argues that evolutionary algorithms should be able to produce human level artificial intelligence as the human brain has biologically evolved. Chalmers then argues that artificial intelligence may be extended to surpass human intelligence as AI may be improved upon. Finally, as more and more sophisticated AI programs are created such programs would be capable of reprogramming to improve themselves in the manner of recursive self-improvement and continue doing so in a rapidly increasing cycle leading to super-intelligence.

It is important to consider what values a super-intelligent entity should be designed to have. It would seem reasonable and desirable to expect such a machine to serve humanity and to:

- It should behave ethically and do good
- Align with human values
- Have a positive effect on humanity
- Contribute to improvement of human species

It is conceivable that super-intelligent machines could pose a threat to humanity, in that they could potentially become so powerful that humans would be unable to control them. There may be unintended consequences to the super-intelligence goals given to these machines, and so the challenge is to build a super-intelligent machine that will serve humanity. The consequence of building a machine that does harm is that such a machine could potentially be so powerful that it could annihilate the human race. And so, it is essential that there are strategies to limit the super-intelligent agent's ability to influence the world.

7.12 Rights of Intelligent Machines

As the intelligence of machines evolve it is reasonable to ask how machines should be treated and viewed in society? If machines evolve to the extent to which they have emotion and empathy will they be considered as animals, as humans, or as inanimate objects? The question on what rights (if any) should be given to robots being robots, what rights should be given as robots evolve, and how should such rights be protected.

It seems inappropriate to give human rights to robots at this time, until at least, they are really indistinguishable from humans. Robots are pure machines in that they are inanimate objects like cars, and they are not sentient creatures with empathy and emotions. Every human being is a unique creation of its parents and society, and every human has a unique sense of self. That is, every human is the outcome of their social condition of family, education, experience and relationships.

Robots have the ability to simulate or emulate human intelligence, and are significantly faster in performing scientific calculations than humans. However, they lack important human characteristics such as consciousness, awareness of self, and empathy for others, and while every human being is a unique creation a robot may be replicated and mass-produced (Fig. 7.7).

Fig. 7.7 TOSY Ping Pong playing robot 2009

The question thus seems to be what are the appropriate rights to be given to robots at this time and in the evolving future, and as robots are part of the modern world, society has a responsibility to the welfare of robots and sophisticated modern technology. Robots have a right to be well designed, and to be built to a high quality so that they are safe for the public to use. They have a right to be built to be trustworthy to wider society, and a right not to be subjected to any cruelty or misuse by humans. They have a right to be repaired (subject to cost feasibility) so that identified problems may be resolved when they malfunction. They have a right to be disposed of in an environmentally sustainable manner.

These are the initial rights at this time, and are likely to evolve over time as robots become more and more sophisticated. The rights to be given to super-intelligent machines will be significantly greater than these.

7.13 ChatGPT

ChatGPT (Chat Generative Pre-trained Transformer) is a large language model based AI chatbot developed by OpenAI that was launched in late 2022. The initial version was text-based, but it has since been updated to handle audio and visuals. The chatbot is able to generate impressive "human-like" text, and the language model is able to respond to questions, as well as composing articles and essays, code and social media posts. ChatGPT is a form of generative AI, which is an AI tool that allows the user to enter a question, and the tool generates human-like text and images.

Users can ask ChatGPT a variety of questions including both simple and complex questions, and the tool is also able to write and debug code. The user can also ask it for more information or ask it to try again. People have used ChatGPT to do various tasks such as:

- Writing code and detecting defects
- Composing music
- Drafting emails
- Creating articles for websites.

ChatGPT is trained to generate text based on input, and it does not understand the complexities of human language. Instead, it is a statistical engine for pattern matching that uses terabytes of data, and extrapolates the most probable answer to a scientific question. This involves taking huge amounts of data and searching for patterns in the data, and the tool becomes increasing proficient at generating humanlike language and thought.

Machine learning is concerned with description and prediction, and does not employ causal reasoning or physical laws. Human-style thought is based on possible explanations with error correcting, and this is a process that gradually limits what possibilities may be rationally considered. ChatGPT is unlimited in what it may learn (memorise), but it is incapable of distinguishing the possible from the impossible, as these just represent different probabilities that could change over time. This means that the predictions of ChatGPT need to be treated with scepticism and caution [7].

There are ethical concerns with ChatGPT such as the dangers of cheating and plagiarism in the education sector, where students use ChatGPT to write essays or papers. ChatGPT may produce inaccurate information and may be used to spread misinformation. ChatGPT can write code, and so it could potentially write malicious software that could cause harm. ChatGPT could be trained to copy a person's writing style, and so could be used to impersonate another person. There may be a bias in the training data leading to inaccurate output, and there are potential privacy issues with the technology.

7.14 Risks and Issues with AI

We have discussed some of the benefits and potential of the AI field, and in this section we summarise some of the risks /issues of the field (Table 7.6).

Table 7.6 Risks and issues with AI

Risk No	Description
1	How can we prevent AI from posing threats to Human Dignity?
2	How can we prevent AI from posing threats to privacy?
3	AI could be abused by an oppressive state to control its population with mass surveillance of its citizens and facial recognition software. How can this be prevented?
4	It is essential to encode an appropriate code of ethics into an artificial device. How can this be done? How can it be verified? How should abuses of the code of ethics be handled?
5	Who should be held accountable if an AI device makes an incorrect decision?
6	Biased algorithms may lead to discrimination. Who should be held accountable if this happens? What avenues are open to those negatively affected by such algorithms?
7	How can we prevent Deepfake technology from harming others?
8	Should autonomous weapon systems be banned? What regulations/safeguards are needed to protect the public?
9	Super-intelligent machines could pose threats to the survival of the human race. Should research on super-intelligent machines be banned? How can the human race be protected?
10	How can we encode consciousness into AI machines?
11	How can we trust an AI machine?
12	How can we know if an AI machine is ethical?
13	What rights should be given to super-intelligent machines?
14	Should we exploit super-intelligent machines?

7.15 Review Questions

1. What is AI?
2. Explain the significance of Weizenbaum's Eliza program to privacy in AI.
3. What are the challenges of self-driving vehicles?
4. Explain the significance of the trolley problem to autonomous vehicles.
5. How should ethics be taught to an autonomous vehicle?
6. What are the ethical issues that could arise with expert systems?
7. What are the potential impacts of AI on unemployment?
8. Explain how AI algorithms may lead to discrimination and racial profiling.
9. What are the rights of intelligent machines?

7.16 Summary

The goal of Artificial Intelligence is to create a thinking machine that is intelligent. It is a young multi-disciplinary field with branches including logic, philosophy; psychology; linguistics; machine vision; and expert systems.

Alan Turing devised the Turing Test as a way to test the intelligent behaviour of a machine, and Searle devised a rebuttal demonstrating that even if a machine passes the Turing test it may not be considered intelligent.

There are philosophical and moral problems to consider in AI such as the exploitation of artificial machines by humans, and whether it is ethical to do this. Weizenbaum created the famous Eliza program at MIT in the mid-1960s, and he was shocked to discover that so many users were convinced that the program had real understanding.

A driverless car is a vehicle that can sense its environment and navigate its way without human intervention. Driverless cars will need to be encoded with a moral compass to deal with situations where ethical decisions needs to be made.

An expert system is a computer system that attempts to copy the reasoning process of human experts in a narrow specialised domain. It consists of a set of rules (or knowledge) supplied by the domain experts about a specific class of problems, and it allows knowledge to be stored and intelligently retrieved.

AI and facial recognition technology has facilitated the recognition of individuals at mass gatherings, and there is a danger that such technologies could facilitate the development of a totalitarian state that uses these technologies to control its population and to eliminate those that pose a threat to the state.

References

1. A. Turing, Computing, machinery and intelligence. Mind **49**, 433–460 (1950)
2. G. O' Regan, *A Brief History of Computing*, 3rd edn (Springer, Heidelberg, 2021)
3. J. Weizenbaum, Eliza, A computer program for the study of natural language. Communication between man and machine. Commun. ACM **9**(1) 36–45 (1966)
4. Joseph, *Computer Power and Human Reason: From Judgments to Calculation*
5. Can we teach robots ethics. *BBC Magazine*, 17th Oct 2017
6. D. Chalmers, The singularity. A philosophical analysis. J. Consciousness Stud. **17**, 7–65 (2010)
7. N. Chomsky, *The False Promise of ChatGPT*, 8th Mar 2023, New York Times

Introduction to Law

8

Abstract

This chapter provides a brief introduction to law, where the laws of a country apply to all citizens and residents of the state. Roman law influenced the development of civil law, where the civil law tradition is focused on reasoning on the basis of rules and is the dominant tradition in Western Europe. Sharia law consists of a set of duties that all Muslims are expected to observe, and the law acts as a path to guide Muslims in their relationships with their neighbours, the state and with God. English common law operates on the principle of binding precedent, where the judge in a particular case must follow the decision of judges that have ruled on similar cases in the past. We discuss European law and the European convention of human rights.

Keywords

English Common Law · Magna Carta · Civil Law · Sharia Law · Roman Law

8.1 Introduction

Modern society is governed by various rules of behaviour such as rules of etiquette, rules from specific religious traditions, rules of membership of an organisation, and the laws of the state (e.g., legal rules such as obeying the rules of the road). The laws are generally made by the legislature of the state, and they apply to all citizens and residents in the state. The laws may be monitored and enforced by the state, with sanctions such as fines or imprisonment imposed on offenders when the laws are not followed.

The origins of law are as old as the cradle of civilisation itself, with Babyloynian law developed by King Hammurabi in Mesopotamia c.1750 B.C. This code awarded harsh penalties to the perpetrators of crime, with physical punishment (e.g., removal of the guilty person's tongue, hand or ear) applied for specific crimes. It is one of the earliest legal codes where it is presumed that the accused person is innocent until proven guilty.

Draco introduced the Draconian law code in Athens c. 620 B.C., and some of these laws were harsh and severe (leading to the term "draconian" to describe such laws). For example, any debtor whose status was lower than that of his creditor was forced into slavery, and the death penalty was applied for relatively minor offences. Solon reformed Draco's laws in the sixth century B.C., and all debts were cancelled, and those who were enslaved due to debt freed. Further, borrowing was prohibited where the collateral for the loan was the individual's freedom. Solon also reformed the Athenian constitution, and Solon's legal code replaced Draco's harsh laws.

The Athenian legal system used courts to settle disputes between people, and the jury would decide if the accused was guilty, and, if so, on the appropriate punishment to apply. There were no professional lawyers employed as such, but there were teachers of rhetoric (the sophists) in Athens who were skilled in teaching the art of argument and persuasion. The jury members were ordinary members of Athenian society, and could consist of 200–400 members for private or public law cases such as the trial of Socrates as described in Plato's Apology.

Roman law influenced the development of civil law, which is the most widely used legal system today. Roman law spanned a period of almost 1000 years from the early republic to the late Empire. The earliest Roman civil law c. 450 B.C. was written down in the Twelve Tables, and developments continued up to the legal code of the Emperor Justinian (*Corpus Juris Civilis*). The latter was developed in the Eastern Roman Empire c. 530 A.D., and it was an attempt to codify the existing Roman Law. It consisted of the *Codex* (imperial legislation), the *Digest* (writings of the jurists such as magistrates who had deep expertise on law), and the *Institutions* (a student textbook). The Western Roman Empire came to an end in the late fifth century A.D., but Roman law remained in use in the Eastern Roman Empire up to the fall of Constantinople in 1453 A.D. Roman law was interpreted and adapted for use in western Europe after the fall of Rome, and it forms the basis of the legal codes of many countries in western Europe.

Roman law developed from the early republic with *jus civile* (civil law) the body of common laws related to Roman citizens and based on customs and legislation; *jus gentium* (laws of nations) was developed by magistrates and governors to apply equally to Roman citizens and foreigners. The Romans divided the law into *jus scriptum* (written law such as legislation) and *jus non scriptum* (unwritten law such as customs).

Islam was developed by the prophet Mohammed in the Arabian Peninsula in the seventh century A.D., and the earliest Muslims consisted of various Arab tribes from Mecca and Medina. *Sharia* Law (the pathway to be followed) consists of a set of duties that all Muslims are expected to observe, and the law acts as a path to guide Muslims in their relationships with their neighbours, the state and with God. Sharia law plays an important role in administering justice in several countries in

8.1 Introduction

the Middle East, as well as playing a role in parts of the legal system in some Muslim countries in Africa and Asia (see Sect. 9.4).

English common law developed in England from the twelfth century, when King Henry II developed a single system of justice for the entire country that would be under the control of the king. Judges play an important role in making the law in the common law system, as their judgments establish legal principles. The system operates on the principle of *binding precedent*, where the judge in a particular case must follow the decision of judges that have ruled on similar cases in the past (i.e., the judge needs to be consistent with previous judgements for similar cases).

The origin of *civil law* is from the Roman world, and this is a codified system that specifies what may be brought to court as well as the applicable procedures and punishment. Codified laws refer to the *rules and regulations that have been written down for the purpose of providing civil order in a society*. These laws are generally produced by legislation in parliaments, and judges interpret the law and the intentions of parliament. They may interpret the law literally or modify the interpretation (e.g., extending the definition in a statue or considering what problem the legislators were attempting to solve) to prevent absurd results. The role of the judge is to establish the facts and to apply the applicable code. The main systems of law used around the world are summarised in Table 8.1.

There are various types of law including

- Constitutional Law
- Criminal Law
- Administrative Law
- Property Law
- Contract Law
- Family Law
- Employment Law
- Natural Law
- Human rights Law
- European Law
- International Law.

The distinction between civil law and criminal law is that *civil law* involves the resolution of disputes between individuals, or between individuals and businesses, whereas *criminal law* is the prosecution of an individual by the state for alleged wrongdoing. There are different burdens of proof in civil and criminal cases, with criminal law generally requiring a high burden of proof (beyond reasonable doubt), as the consequences of a successful prosecution by the state could lead to imprisonment and the deprivation of the accused person's liberty.

The accused person has a right to a fair trial in an impartial court, the right to make a defence, and due process is a fundamental principle in the legal system, which involves respecting the rights of the accused and fair treatment throughout the judicial process.

Table 8.1 Systems of Law

System of Law	Description
English Common Law	This system of law is used in England, Australia, New Zealand, the United States and other countries Legal principles in decisions of judges Operates on the principle of binding precedent (judge in particular case must follow the decision of judges in previous similar cases) May not be a written system of codified laws Decisions are binding Decisions of highest court may only be overturned by the same court or through legislation Everything is permitted that is not prohibited by law
Civil Law	Origin is in Roman law It is a codified system Comprehensive legal codes that specify all matters that may be brought to court, as well as the applicable procedure and punishment Judge's role is to establish facts and apply provisions of applicable code Judges decision in framing law much less compared to legislators/legal scholars who draft the code Written constitutions based on specific codes
Mixed Systems	Some elements of English common law and some elements of Civil Law
Sharia Law	This system of law with its origins in desert tribal society on the Arabian peninsula, and it is strictly applied in some Muslim states
Canon Law	The system of law of the Roman Catholic Church, and it deals with internal organisation of the church and civil affairs such as marriages, contracts and wills
Natural Law	The system of law based on the values in human nature that are applied in the law of the state. Natural law argues that people have inherent or natural rights conferred on them by God or nature This includes law whose content is established purely by means of reasoning
Brehon Law	This refers to early Irish law in medieval Ireland with parts of it written down in the sixth century. Brehon law recognised divorce and equality of the genders. The system was largely self-enforcing

8.2 English Common Law

William the conqueror of Normandy invaded England in 1066 and defeated the English (the Anglo Saxon, King Harold II) at the Battle of Hastings to become the first Norman king of England. At that time, justice was administered according to the different customs and traditions of various part of England, and William began a process towards centralised law. He set up a king's court (curia regis) to begin, and as he travelled around the country citizens would bring their grievances to him. The king or his advisors would consider the grievance and make a judgment on the dispute.

8.2 English Common Law

King Henry II was the first Plantaganet[1] king of England (House of Anjou in France), and he was the great grandson of William the conqueror and the father of Richard the Lionheart and King John. King Henry established a single system of justice for the entire country in 1154 that would be under control of the king, and local laws were replaced over time by new national laws that would be common to all citizens (i.e., *common law*).

Justices travelled around the country to administer justice, and in time the decisions of these judges were written down and published. This led to past legal decisions made by judges to be argued for in similar cases before the courts. This led to the principle of *binding precedent*, where a previous judgment is binding on future cases, and so the fact that the law develops from previously decided cases is a central feature of common law.

The *Magna Carta* (great charter) was introduced in 1215 during the reign of King John, and it became a symbol of liberty and human rights that was to later influence the US Declaration of Independence and the UN Declaration of Human Rights. The motivation for its development was the concerns that the barons of England had with King John and on his use of power, and the charter was designed to limit the power of the monarch and to give protection to free men.

The charter applied only to freemen, and so it did not include the peasants who were ruled over by their landowner. Among the important rights and principles in the charter is that "*All free men have a right to justice and a fair trial*", and that everyone, including rulers and leaders, must obey the law. The Magna Carta became part of English Common Law, and it was reissued several times. However, it is a document of its time and says nothing on free speech, personal privacy, religious freedom and peaceful protest (Fig. 8.1).

The *principle of equity* was introduced in common law in the fifteenth century, and it refers to a set of remedies and associated procedures with civil law. This principle of law is focused on obtaining a fair result, and it is important where more than a legal solution of compensation is required, and where specific action is required (e.g., an injunction) to achieve a just solution. The chancery courts (i.e., courts of equity) were established in parallel with common law courts, and these courts considered the circumstances of the case with the goal of imposing a fair and equitable remedy to an injustice.

The solution to an injustice could take the form of an injunction, which could be a judicial order restraining a person from performing a certain action, or compelling the individual to carry out a specific action. The courts of common law and courts of equity were combined in 1873.

The *Bill of rights Act* of 1689 was introduced in England shortly after the Glorious revolution of 1688–89. The revolution led to the overthrow of the Catholic King James II, and William III and Mary II replaced him. It is an important part

---

[1] The Plantaganets remained kings of England for over 300 years, from 1154 (following the accession of Henry II to the throne) to 1485 (the death of Richard III at the battle of Bosworth during the War of the Roses).

Fig. 8.1 King John signs Magna Carta

of constitutional law in England, and it placed further limits on the power of the monarch, and ultimately gave parliament power over the monarchy. It set out the rights of parliament including free speech, free elections and certain fundamental rights and liberties of individuals. Many of the ideas in the act reflected the views of the philosopher, John Locke. The bill of rights influenced the development of civil liberties and human rights, including the US Bill of Rights, the UN Universal Declaration of Human Rights and the *European Convention on Human Rights*. It insists on due process in criminal trials, and prohibits cruel and illegal punishments.

The constitution of a country is a set of laws, rules and practices that create the institutions of the state, including the powers of the institutions, the relationship between them, and the relationship between the institutions and the citizen. The British constitution is monarchical and uncodified: i.e., there is no single document that systematically describes the laws and institutions of the British state. Instead, there is a collection of documents that describe the sources of the British constitution, including the sovereignty of parliament, the acts of parliament, the Magna Carta and the Bill of Rights Act of 1689.

The sources of British law include the *statues* (or acts of parliament), common law, European and International Law, and the European convention on Human rights.[2] The principle of judicial precedent is very important in common law as a way for making laws, but in modern times the laws made by parliament have become the most significant means of making law. *Royal assent* is the method by which the monarch formally approves legislation thereby making the bill an act of parliament.

A key principle of the British constitution is that of the *supremacy of parliament*, where parliament may enact, amend or revoke any law that it sees fit. The Human Rights Act of 1998 recognises that parliament has the 'power to make legislation that is incompatible with the rights enshrined in the European convention for human rights, but section 3 of the Human Rights Act means that UK courts now interpret all statues with the rights granted under the European convention for Human Rights. However, if a UK statue is incompatible with an article in the European convention then a British court can declare that incompatibility (under section 4 of the act) and advise parliament, but the court cannot force a change in the law. That is, a UK court cannot force parliament to change the law: i.e., *parliament is sovereign* and it is supreme over all other institutions of the state.

However, political pressure following such a declaration could result in an amendment of the affected law. This led the UK to introduce the Human Rights Act of 1998 to domesticate the rights enshrined in the European Convention of Human Rights, and as these rights are now part of British Law they may be considered before the court. And so, in a sense the introduction of this act brought human rights back home, and this was a balance between protecting fundamental human rights and the sovereignty of the UK parliament.

Laws made in other countries (e.g., in the Republic of Ireland) need to be compatible with their constitution (which is the highest law in the state, and it represents the key principles and values of the state, as well as how the state is to be governed). Further, a law that is ruled to be incompatible with the constitution by the Supreme Court in the state is said to be unconstitutional, and it needs to be amended by the legislature of the state (i.e., the Dáil and Seanad).

The principle of the *separation of powers* is an integral part of constitutional law, where the three branches of government (the legislative, executive and judiciary) are kept separate. The goal is to avoid an over-concentration of power in one branch of government and to provide checks and balances (there is truth in the well known proverb that absolute power corrupts absolutely). The idea of mixed government was mentioned by Aristotle in the fourth century B.C., and the French enlightenment philosopher, Montesquieu, discussed the distribution of power of the legislative, executive and judiciary in "The Spirit of Laws" [1], which became known as the *tripartite system*. The separation of powers in the US constitution is illustrated in Fig. 8.2.

---

[2] The UK left the European Union on Dec 31st 2020 and so it is unclear at this time as to what extent (if any) European law will play a role in British law in the future.

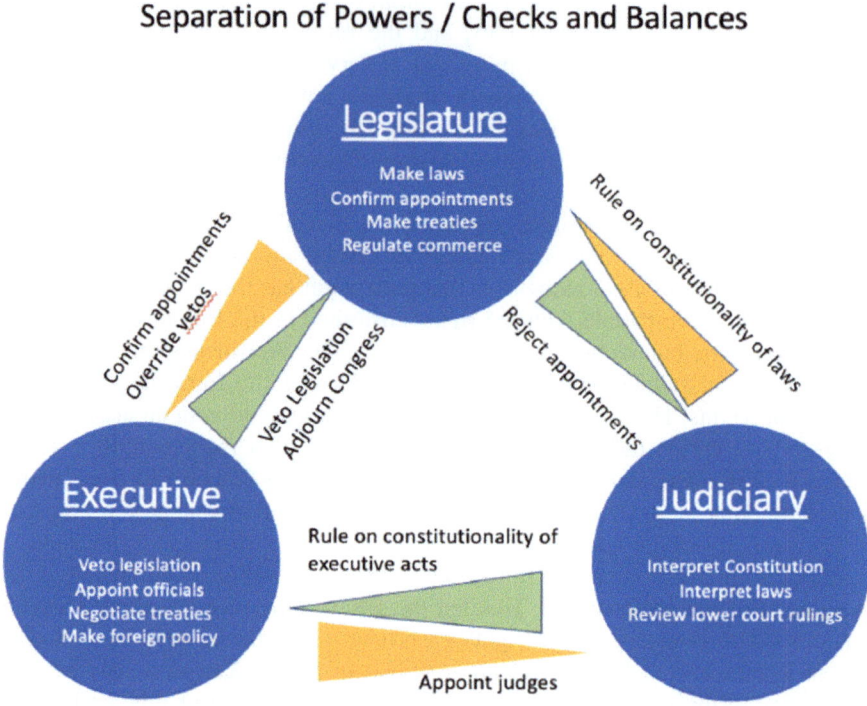

Fig. 8.2 Separation of powers in US Constitution

The British system is not fully separate, as the legislative and executive intersect (to become a minister in the executive one must be a member of parliament or the House of Lords). The judiciary is independent and the Supreme Court was established as the court of final appeal in 2009, as a replacement for the House of Lords Judicial Committee.

There are several principles underlying common law and justice in England, and it is fundamental that if there is evidence given against a party then that party is given time to consider the evidence. There is no punishment without breach of the law,[3] and the judge is a neutral party who listens to both sides and considers the evidence in the dispute between both parties before making a rational and fair judgment. The judiciary are independent and are free to exercise their judicial powers without interference from the litigants. The key legal principles underlying English common law include:

---

[3] We shall not consider the serious miscarriages of justice that occurred during the IRA bombing campaign in Britain in the 1970s, when several innocent people were wrongly convicted of committing atrocities. These include the infamous cases of the Birmingham six and the Guildford four, where ten innocent people were wrongly convicted of the bombings of bars in Birmingham and Guildford, and spent many years in prison until their convictions were overturned.

- Independent judiciary
- Fairness in the proceedings
- Due process of legal rights of person
- Judge is an unbiased and neutral party
- Judge hears both sides before making judgment
- Proper procedures of law followed
- Equality before the law
- Supremacy of the regular law
- No punishment without breach of law.

Case law refers to the creation of law by the decisions of the judges in court, and English common law uses case law to a large extent as a source for legal rules and principles. A *criminal case* arises after the police arrest and charge a person with an offence, and the crown prosecution service (or the Director of Public Prosecutions or DPP in Ireland) prosecutes the case in the criminal courts on behalf of the state.

A *civil case* arises when an individual or organisation takes legal action against another individual or organisation (e.g., a for breach of contract). When these legal cases are decided by judges in the courts then these decisions make up case law, and it is usually only the decisions and legal principles employed of the higher courts (such as the high court and supreme court) that become part of case law.

The concept of *judicial precedence* is at the heart of case law, and it essentially states that *like cases should be treated alike*. That is, when judges decide a case they will consider whether a similar case has come before the courts to decide the current case in a similar way to ensure consistency of judgments. The doctrine of judicial precedence is binding in that the judge must follow earlier court decisions from higher courts in respect to the legal rules that the decision was based upon (i.e., the legal rationale for the decision).

The principle of judicial precedence helps to ensure consistency in judgments, and that the judgment is based on a consistent set of formal rules. There is a tension between consistency and flexibility, as significant societal changes over time could lead to outdated law that does not reflect the realities of modern society. That is, too rigid an adherence to precedence could lead to an injustice in a particular case, and impede the proper development of the law.

8.3 Civil Law

The civil law tradition is dominant in most of Western Europe, and is focused on reasoning on the basis of rules. Following the collapse of the Western Roman Empire in the late fifth century, Europe was divided into a number of territories with diverse customs and laws, although some basic remnants of Roman law remained. Roman Law (the Digest) was rediscovered in northern Italy in the eleventh century, and the historical Roman Digest was the writings of the jurists in the sixth century as discussed earlier in the chapter. The Roman Digest influenced the study and development of the law in continental Europe, and it was studied in the early universities.

Further, the Christian church had grown in importance from Roman times, and Canon Law (the law of the Church), became important as it dealt with the law of the church, as well as civil matters such as marriages and wills. In practice, Roman law and Canon law were usually studied together, and over time this new *ius commune* (or European common law) became more influential. Roman law eventually became dominant in Europe, as it was seen to be a rational system with good rules, and the laws of European countries have been greatly influenced from the combination of Roman and Canon laws.

The French revolution led to the codification of French law under the Emperor Napoleon, and the codes led to consistency in the application of the law in various parts of France. It created legal certainty for citizens as the law was written down and could be examined by everyone, and it emphasised the powers of the legislature in making laws (as distinct from the way that judges make the law under English common law). Further, it guaranteed that people could influence the content of the law through the democratic process.

The Napoleonic campaign led to the spread of French thinking on codification to various parts of Europe, and the French approach spread to other European countries such as Belgium, Spain and Portugal. The approach in Germany (in Prussia) was historical with a study of the Roman Digest by legal scholars, and they adapted the law to meet the needs of society. The German approach spread to Austria and Switzerland, whereas some other European countries such as Italy, Poland and Holland were influenced by both the French and German traditions.

The development of law in the twentieth century includes the development of human rights law after the Second World War to deal with the horrors of the Holocaust and the war crimes committed by the Wehrmacht. The creation of the European Economic Community (EEC) and its later evolution into the European Union led to a vast collection of supranational laws for the member states, as well as various treaties, conventions, rules and regulations that govern international trade.

The key feature of the civil law tradition is that it is focused on reasoning on the basis of rules. The rules are supplemented with canons of interpretation that are developed to enable legal decision makers to use the rules and to apply them to concrete cases. The interpretation is said to be *literal* (or grammatical) if the legal decision maker interprets the rule literally, whereas if the interpretation takes the *legislative intent* (intention of the legislator in solving the problem) into account then the legal decision maker is said to apply the *Mischief Rule* (or Legislative intent). The Golden rule is when the legal decision maker tries to determine the purpose of the rule himself (or herself).

Arbitration is a way to resolve commercial disputes, where a person who is not an official judge makes a decision that is accepted by the parties.

8.4 Sharia Law

Sharia law arose on the Arabian Peninsula and consists of a set of duties that all Muslims are expected to observe. The law acts as a path to guide Muslims in their relationships with their neighbours, the state and with God.

The five pillars of Islam include the profession of faith (*shahada*), daily prayers (*salat*), almsgiving (*zakat*), fasting from sunrise to sunset during Ramadan (*sawn*), and a pilgrimage at least once in a person's lifetime to Mecca (*hajj*).

Sharia indicates what an individual may or may not do in law, and also what the individual should or should not do in conscience. That is, sharia law is more than a system of law, but is also a code of behaviour that guides the devout Muslim in both their private and public activities (Fig. 8.3).

Muslims regard Sharia Law as revelation in the expression of the Divine will as revealed to the Prophet, and Sharia law is viewed as an unchanging continuity (especially for the Islamic rituals part of Sharia). The expansion of Islam to other countries led to some divergence in the legal traditions as well as different schools of Islam. Some of the schools disagreed with respect to the reliance that they placed on the *Hadith* (the sayings of the prophet Mohammed), which are believed to be an authentic account by other scholars. Some modern scholars believe that many of the sayings in the *Hadith* were written by others, and fictitiously ascribed to the Prophet in order to give their doctrines greater authority.

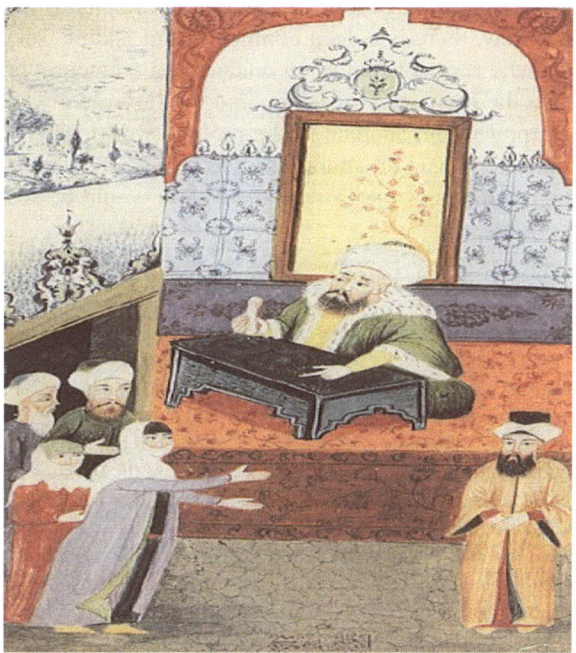

Fig. 8.3 Sharia family law court case

Sharia law covers offences against another person such as homicide or assault, which are punishable by retaliation with the offender subjected to the same treatment as the victim. The victim's family may accept compensation or blood money (*diyah*) instead. The punishment is fixed (*hadd*) for certain crimes: for *apostasy* (renunciation of religious faith) the penalty is death; the penalty for theft is amputation of the hand; death by stoning for extramarital affairs where the offender is married, and 100 lashes where the offender is unmarried; and 80 lashes for consuming alcohol. Family law is patriarchal and reflects the nature of Arabian tribal society in the early years of Islam, with fathers having the right to contract their daughters in marriage, and husbands having the right of polygamy with up to four wives.[4]

The Golden Age of Islamic civilization was from 750 A.D. to 1250 A.D., and during this period enlightened caliphs recognised the value of knowledge, and sponsored scholars to come to Baghdad to gather and translate the existing world knowledge into Arabic. This led to the preservation of western knowledge during the Dark ages in Europe. Further, the Islamic cities of Baghdad, Cordoba and Cairo became key intellectual centres, and scholars added to existing knowledge (e.g., in mathematics, astronomy, medicine and philosophy), as well as translating the known knowledge into Arabic. Education was important during the Golden Age, and the Al Azhar University in Cairo was established in 970 A.D. The Islamic World has created beautiful architecture and art including the ninth century Great Mosque of Samarra in Iraq; the tenth century Great Mosque of Cordoba in Spain; and the eleventh century Alhambra in Granada (Fig. 8.4).

Today, there is a deep conflict within Islamic societies between the traditionalists who regard Islam as an unchanging collection of truths that are as valid today as in the early days of Islam, and the reformers who argue that Islam needs to be interpreted and adapted to the modern world. There has been a growth in Islamic fundamentalism an attempt by Isis to set up a caliphate in the Middle East, where the teachings of the Qur'an and Sharia would be strictly observed.

---

[4] The author recalls a barbaric incident while residing in Saudi Arabia where a poor Sri Lanken maid (Rizana Nafeek) was executed (decapitation) in 2013, as she was held responsible for the death of a 4-month year old child who died in her care in 2005. She was hired as a general maid and had no prior training in childcare, and she made a confession (under duress) at the police station (in theory, Sharia law does not accept statements made under duress). She was tried without legal representation (legal representation was granted for her appeal), and she was a minor at the time of the incident (Saudi Arabia is a signatory to the Convention on the Rights of the Child and is obliged not to execute a minor). Unfortunately, many of the expatriates in the Gulf countries are from poor backgrounds from India, Sri Lanka, Pakistan and the Philippines, and these domestic workers are often treated very badly by their employers, with many working 15+ hours a day with little time off, and the Saudi visa system requires the worker to be sponsored by the employer (*kafeel*) and makes it extremely difficult to change employment. The term "*inshallah*" (God Willing) is a very popular expression in the Middle East, but at times it seems like a fatalistic interpretation of the status quo, where the individual is not playing their part in shaping the world to create a fair and just society.

Fig. 8.4 Al Azhar University. Cairo

8.5 European Law

The origin of the European Union is in the aftermath of the devastation and massive loss of life of the Second World War, and the determination of European leaders (especially of France and Germany) to develop closer cooperation and economic integration between European countries to ensure that war would never return again to the continent.[5] The European Coal and Steel Community (ECSC) was formed in 1951, and its goal was to develop a common market in coal and steel. This led to the formation of the European Economic Community (EEC) with the Treaty of Rome in 1957 with the same six members of the ECSC, and the membership of the EEC expanded over time to nine countries when the UK, Ireland and Denmark joined in 1973. Further expansion of the EEC has taken place since then and today the European Union (EU) is the successor of the EEC and is a supranational organisation consisting of twenty seven member states.

The Treaty of the European Union (TEU) or Maastricht Treaty was ratified in 1993, and this is the formation treaty of the European Union (EU), which is the successor to the EEC. The TEU is combined with the Treaty on the Functioning of the European Union (TFEU) to form the constitutional basis of the European Union. The TEU and TFEU are the main sources of law in the European Union (Fig. 8.5).

---

[5] The English/Irish Quaker, William Penn, had proposed a European council/parliament back in 1693 that would create lasting peace and security in Europe. Penn wrote several articles on religion and tolerance, and he was the founder of the state of Pennsylvania.

Fig. 8.5 Flag of European Union

The European Union is neither a state nor a federation, and it is a supranational organisation with its own institutions and law making powers. The EU powers are conferred upon it from the member states via the various EU treaties, and so the EU exercises *conferred powers*. The treaties established a number of institutions in the EU such as:

- European Parliament
- European Commission
- European Council
- Council of the European Union
- Courts of Justice
- Court of Auditors
- European Central Bank.

European Union law takes precedence over the legal system of member states, and so if there is a conflict between national law of the state and EU law then EU law is supreme. Further, EU citizens can rely on the provisions of EU law in their national courts, and so this ensures the effectiveness of the implementation of EU law in member states. EU regulations and directives place obligations on member states, and these need to be implemented appropriately in each member state.

The primacy of EU law impacted the perceived sovereignty of the British parliament, and sovereignty (as well as English nationalism and immigration) was a key factor in the Brexit referendum in 2016. The 2016 referendum was advisory rather than legally binding (i.e., in theory the government could have ignored the result but politically this was impossible), and it led to the UK's departure from the European Union.

8.6 European Convention on Human Rights

The *Council of Europe* was established in 1949 with the goal of protecting human rights after the slaughter of the Second World War, and to prevent such atrocities occurring again. It initially consisted of ten European countries and now has over 46 signatory states. The *European Convention on Human Rights* (ECHR) was published and came into force in 1953, and it defines various human rights that all signatures to the convention will uphold. Human rights are basic rights and freedoms that belong to every human being, and they are universal and timeless. They include freedom of belief and religion, freedom of expression, freedom to assemble, and freedom to privacy and family life.

The *European Court of Human Rights* applies and protects these fundamental rights and guarantees set out in the Convention, and an individual may take an action against a state where a law of the state breaches the rights granted under the Convention of Human Rights. Further, it is also possible for a state to take action against another state for breaches of rights guaranteed in the Convention.

The court investigates and if it finds that a state's law breaches the rights enshrined in the European Convention then it will impose an obligation on the state to remedy the law. However, if the state does not comply there may be a need to apply political pressure to it, and the state may need to be sanctioned, and in a worse case scenario the state may be expelled from the Council of Europe.

8.7 Natural Rights and Natural Law

Natural rights are rights that are considered to be universal, fundamental and inalienable, and they are not dependent on the laws of a particular country. Natural rights are beyond the authority of a particular government to dismiss, and they are enshrined in international legal treaties. The idea that certain rights are natural and inalienable dates back to the Greeks, where Aristotle and the Stoics considered the concept of natural rights as well as the difference between "just by law" and "just by nature", as well as living in accordance with natural law with virtuous actions.

The Christian theologian and philosopher, St. Thomas Aquinas, argued that there are goods or behaviours that are naturally right or wrong, as God has ordained it that way, and this brings the idea of a divine order into the formulation of natural rights. Later philosophers such as Thomas Paine or Rousseau argued for natural rights on secular grounds using humanist and rationalist theories.

Some philosophers (e.g., Jeremy Bentham and Edmund Burke) have rejected the idea that there are natural rights, and they argue that it is the laws of the state that create rights: i.e., from real laws come real rights, whereas from the laws of nature come imaginary rights. Another words, they argue that natural rights do not exist independently of human endeavour, and they are created by human action, and are the product of a particular society and its legal system.

That is, there is controversy over whether all rights are either natural or legal, with philosophers such as Kant believing that natural rights could be derived by reason alone, and others arguing that rights come from laws. The United States declaration of independence claims to be based on self-evident truths, and states that the creator has endowed citizens with certain inalienable rights such as that all men are created equal.

Natural law (*lex naturalis*) is a system of law based on the values in human nature that are applied in the law of the state. Natural law argues that people have inherent rights conferred on them by God or nature, and the roots of Natural Law are from early Greek and Roman society, and especially in the philosophy of Plato and Aristotle, the Stoics, and in the writings of Cicero.

Cicero gives the classic account of Natural Law in his Republic, where he argues that true law is right reason in agreement with nature, and that it is of universal application, unchanging and everlasting. Cicero states that there is a duty to obey natural law, and that God is its author and enforcer, and that whoever is disobedient to the natural law is denying his human nature and will suffer the worst penalties.

8.8 Human Rights and Human Rights Law

The Universal Declaration of Human Rights (UDHR) in 1948 recognised that all human beings have basic and fundamental rights, and that everyone irrespective of race, religion, ethnicity and gender has these rights. These include civil, political, economic, social and cultural rights, and must be respected and protected in the state, and when they are blatantly disregarded they are violated (Fig. 8.6).

The UDHR declaration was the foundation of International Human Rights Law (IHRL), and international human rights law helps to protect human rights around the world. It lays down obligations that states are obliged to respect, and once states become parties to international treaties they have further duties to protect and fulfil human rights. The state must protect groups and individuals against human rights abuses, and this often involves updating national law to respect and implement protection for human rights. Human rights include:

Fig. 8.6 Eleanor Roosevelt holding the Universal Declaration Human Rights

- Freedom from torture
- Freedom of expression
- Rights to seek asylum
- Right to a fair trial.

The motivation for the establishment of human rights was a response to the atrocities committed during the Second World War. This included the holocaust, where over six million Jews and other minorities, including gypsies, homosexuals, and political dissidents, from various parts of Europe were systematically murdered by the Nazis. Further, the Wehrmacht committed many other war crimes such as massacres, mass rapes, the exploitation of slave labour, and the murder of 2–3 million Soviet prisoners of war during the horrors of the Second World War.

8.9 International Criminal Court

The International Criminal Court (ICC) commenced operations in 2002 as a permanent international court based in the Hague, The Netherlands, with jurisdiction to prosecute individuals for serious international crimes such as:

- Genocide
- Crimes against humanity
- War crimes
- Crimes of aggression.

The idea of an international court to prosecute individuals for serious international crimes was first proposed at the Paris Peace Conference after the First World War in 1919. After the Second World War, the Allied victors established two tribunals to prosecute Axis leaders accused of war crimes, namely the Nuremberg Tribunal, which prosecuted German war leaders, and the Toyko Tribunal, which prosecuted Japanese war leaders.

The UN General Assembly recognised the need for an international criminal court to deal with atrocities in 1948, but it proved to be unrealistic to set up the court during the Cold War. Various tribunals were conducted to prosecute individuals for serious crimes during the break-up of Yugoslavia and the genocide in Rwanda. The international criminal court was finally ratified in 2000.

The ICC is intended to serve as a "court of last resort" in that it acts only if national courts are unable or unwilling to prosecute individuals of serious crimes. It has limited jurisdiction in what it may investigate and prosecute, as it is limited to crimes committed within or by individuals of member states or by a referral from (a deeply divided) United Nations Security Council.

8.10 Freedom of Speech and Responsibility

Freedom of speech refers to the freedom that an individual has to hold and articulate his or her opinions without fear of sanction. The origins of free speech go back to Athenian democracy in the fifth century B.C., and it was introduced into the UK parliament under the 1689 Bill of Rights Act. Freedom of speech includes several rights such as:

- Right to express and disseminate information or ideas
- Right to seek information or ideas
- Right to receive information or ideas
- Right to impart information or ideas.

Freedom of expression is recognised as a fundamental human right in the UN Universal Declaration of Human Rights and in International Human Rights Law, and it covers any media used to obtain or communicate such information. Many countries have protections for free speech enshrined in their constitution, and Article 19 of the UDHR specifies that everyone has the right to hold opinions without interference, as well as the right to seek, receive and impart information irrespective of the medium used. The UDHR contains an important set of principles and rights that while not legally binding are contained in other international treaties that are legally binding.

However, freedom of speech and expression should not be regarded as absolute rights, as the exercise of these rights carry special duties and responsibilities. The limitations or boundaries of free speech are described in Article 19 of the International Covenant on Civil and Political Rights (ICCPR), and are related to slander, libel, incitement, hate speech, copyright violation, and the right to privacy. Mill outlined the importance of introducing limitations on liberty with his *"harm principle"* [2], and so the purpose of these limitations on freedom of expression is to prevent harm to others.

8.11 Review Questions

1. Explain the significance of the Hammurabi code.
2. What are the main systems of law in use today?
3. What is English common law?
4. What is civil law?
5. What is sharia law?
6. What is a good law?
7. What is European law?
8. What is the European convention of human rights?
9. Explain the importance of human rights and human rights law.

8.12 Summary

The laws of a country apply to all citizens and residents of the state, and offenders are subject to sanction such as fines or imprisonment when the laws are broken. The Hammurabi code was developed in ancient Mesopotamia, and it awarded harsh penalties to the perpetrators of crime. It is one of the earliest legal codes where it is presumed that the accused person is innocent until proven guilty. The Athenian legal system used courts to settle disputes between people, and the jury would decide if the accused was guilty, and on the appropriate punishment if so.

Roman law influenced the development of civil law, which is the most widely used legal system today. Sharia law consists of a set of duties that all Muslims are expected to observe, and the law acts as a path to guide Muslims in their relationships with their neighbours, the state and with God. Sharia law plays an important role in administering justice in several countries in the Middle East, as well as playing a role in parts of the legal system in some Muslim countries in Africa and Asia.

English common law developed in England from the twelfth century, and involves judges playing an important role in making the law, as their decisions establish legal principles. It operates on the principle of binding precedent, where the judge in a particular case must follow the decision of judges that have ruled on similar cases in the past.

The civil law tradition is focused on reasoning on the basis of rules and is the dominant tradition in Western Europe. Following the collapse of the Western Roman Empire in the late fifth century, Europe was divided into a number of territories with diverse customs and laws, although some basic remnants of Roman law remained. Roman law eventually became dominant in Europe, as it was seen to be a rational system with good rules.

References

1. B. de Montesquieu, *De L'Esprit des Loix (The Spirit of the Laws)*. Charles de Secondat, France (1750)
2. J.S. Mill, *On Liberty and other Essays*. Oxford World Classics. Originally published in 1859 (1991)

Legal and Ethical Aspects of Outsourcing

9

Abstract

This chapter discusses legal and ethical aspects of outsourcing of software development and/or testing. Advanced economies have many laws and regulations to protect the environment, and the health and safety of employees in the workplace. The laws and regulations of the country where the subcontractor is based may not be as stringent, and so an ethical corporate citizen must do more in that it is not sufficient for the corporation to say that it is complying fully with the laws of every country it is operating, as these laws may not be fit for purpose. Further, the corporation has ethical responsibilities for the health and safety of the subcontractor staff that are working on their projects. We discuss legal aspects of failure including lawsuits and the law of tort.

Keywords

Request for Proposal • Supplier evaluation • Formal Agreement • Statement of Work • Managing supplier • Service Level Agreement • Escrow • Acceptance of Software

9.1 Introduction

Outsourcing is a common business practice where a company contracts out business functions such as manufacturing, software development, and call centres to third party providers. The outsourcing of a business function to a distant country is termed *offshoring*, whereas outsourcing may also be done domestically, and *nearshoring* is the term used where the outsourcing is to a nearby country. The main benefits of outsourcing include:

- Cost savings due to a reduction in business expenses
- Availability of expertise not available in house
- Allows company to focus on its core business activities

- Increased efficiencies.

Outsourcing involves handing control of various business functions over to a third party, and this leads to business risks such as the quality of the service may be below expectations, or the third party may go out of business, or that there may be risks to confidentiality and security of data. Outsourcing involves managing the day-to-day relationship with the offshore/onshore team in possibly different time zones, and there could also be differences in language and culture.

Many large software projects involve total or partial outsourcing of the software development, and it is therefore essential to select a supplier that has the capability and expertise to deliver high-quality and reliable software on time and on budget. Supplier selection and management is concerned with the selection and management of a third-party software supplier. The success of the project is dependent on the expertise and competence of the suppliers, and so it is essential that the selected supplier is competent and capable of delivering high-quality and reliable software on time and on budget.

This means that the process for the selection of the supplier needs to be rigorous, and that the capability of the supplier is clearly understood, and the associated risks are known prior to selection. The selection is based on objective criteria such as cost, the approach proposed, the ability of the supplier to deliver the required solution, the supplier capability, and while cost is an important criterion, it is just one among several other important factors.

Once the selection of the supplier is finalised a legal agreement is drawn up between the contractor and supplier, which states the terms and condition of the contract, as well as the associated statement of work. The statement of work details the work to be carried out, the deliverables to be produced, when they will be produced, the personnel involved as well as their roles and responsibilities, any training that needs to be provided, and the standards to be followed.

The supplier then commences the defined work and is appropriately managed for the duration of the contract. This will involve regular progress reviews, and acceptance testing is carried out prior to accepting the software from the supplier. The following activities are generally employed for supplier selection and management (Table 9.1).

Remote project management is concerned with managing remote and hybrid teams to ensure that the project objectives are achieved. Traditional project management involve teams based in the same physical location, whereas often teams may operate in hybrid mode with some employees working in the office and other employees and teams working remotely in different physical locations. This means that remote employees often play important roles in the success of projects, and remote project management has become important in managing hybrid and remote teams.

The management of remote teams requires modern communication including video conferencing, shared files, and documents, as well as team communication and messaging apps. The creation of the team is the easy part as it is more challenging to build a team culture with remote teams. The project manager will

Table 9.1 Supplier selection and management

Activity	Description
Planning and Requirements	This involves defining the approach to the procurement. It involves: • Defining the procurement requirements • Define evaluation criteria • Form the evaluation team to rate each supplier
Identify Suppliers	This involves identifying suppliers and may involve research, recommendations from colleagues, as well as known suppliers from previous working relationships
Prepare and Issue RFP	This involves the preparation and issuing of the Request for Proposal (RFP) to potential suppliers. The RFP may include the evaluation criteria and a preliminary legal agreement
Evaluate Proposals	The received proposals are evaluated and a short-list of three to five potential suppliers produced. The short-listed suppliers are invited to make a presentation of their proposed solution
Select Supplier	Each short-listed supplier makes a presentation followed by a Q&A session. The evaluation criteria are completed for each supplier and reference sites checked (as appropriate). The decision on the preferred supplier is made
Define Supplier Agreement	A formal agreement is made with the preferred supplier. This may include • Negotiations with the supplier/involvement with Legal Department • Agreement may vary (Statement of Work, Service Level Agreement, Escrow, etc.) • Formal Agreement signed by both parties • Unsuccessful parties informed • Purchase Order raised
Managing the supplier	This is concerned with monitoring progress, project risks, milestones and issues, and taking action when progress deviates from expectations
Acceptance	This is concerned with the acceptance of the software, and involves acceptance testing to ensure that the supplied software is fit for purpose
Rollout	This is concerned with the deployment of the software and support/maintenance activities

stay engaged with the team throughout the project with virtual meetings, and remote project management is like traditional project management except that the project is executed remotely. It is a flexible methodology that can support various approaches such as traditional software engineering and Agile.

The project manager needs to determine the remote structure that is required, and the project expectations need to be communicated clearly to the team members (including the outsourced team members) at project initiation, including the process to be followed, work hours, project goals, their responsibilities, the tools that will be employed for collaboration, and so on. The project manager will define and manage the expectations of each team member, and will conduct regular (virtual) team meetings during the project. The project manager will check in regularly (daily) with team members to determine progress made with respect to their responsibilities, and update the project schedule accordingly.

9.2 Planning and Requirements

The potential acquisition of software arises as part of a make-or-buy analysis at project initiation. The decision is whether the existing project team should (or has the competence to) develop a particular software system (or component of it), or whether there is a need to outsource (or purchase off-the-shelf) part or all of the required software. The supplied software may be the complete solution to the project's requirements, or it may need to be integrated with other software produced for the project. The following tasks are involved:

The requirements are defined (these may be a subset of the overall business requirements).
The solution may be available as an off-the-shelf or open source software package (with configuration needed to meet the requirements).
The solution may be to outsource all or part of the software development.
The solution may be a combination of the above.

Once the decision has been made to outsource or purchase an off-the-shelf solution an evaluation team is formed to identify potential suppliers, and evaluation criteria is defined to enable each supplier's solution to be objectively rated.

A plan will be prepared by the project manager detailing the approach to the procurement, defining how the evaluation will be conducted, defining the members of the evaluation team and their roles and responsibilities, and preparing a schedule of the procurement activities to be carried out.

The next part of this chapter is focused on the selection of a supplier for the outsourcing of all (or part) of the software development, but it could be adapted to deal with the selection of an off-the-shelf software package.

9.3 Identifying Suppliers

A list of potential suppliers may be determined in various ways including:

Previous working relationship with suppliers
Research via the Internet/Gartner
Recommendations from colleagues or another company
Advertisements/other.

A previous working relationship with a supplier provides useful information on the capability of the supplier, and whether it would be a suitable candidate for the work to be done. Further, a supplier that is ISO 9001 certified for quality and ISO 27001 certified for Information Security has independent indications of reasonable capability. Companies will often maintain a list of preferred suppliers, and these are the suppliers that have worked previously with the company, and whose capability is known. The risks associated with a supplier on the preferred supplier list are known, and they are generally less than those of an unknown supplier. If the experience of working with the supplier is poor, then the supplier may be removed from the preferred supplier list.

There may be additional requirements for public procurement to ensure fairness in the procurement process, and often-public contracts need to be more widely advertised to allow all interested parties the opportunity to make a proposal to provide the product or service.

The list of candidate suppliers may potentially be quite large, and so short listing may be employed to reduce the list to a more manageable size.

9.4 Prepare and Issue RFP

The Request for Proposal (RFP) is prepared and issued to potential suppliers, and the suppliers are required to complete a proposal detailing the solution that they will provide, as well as the associated costs, by the closing date. The proposal will need to detail the specifics of the supplier's solution, and it needs to show how the supplier plans to implement the requirements.

The RFP details the requirements for the software, and must contain sufficient information to allow the candidate supplier to provide a complete and accurate response. The completed proposal will include technical and financial information, and will contain sufficient detail to allow a rigorous evaluation to be carried out.

The RFP may include the criteria defined to evaluate the supplier, and often weightings are employed to reflect the importance of individual criteria. The evaluation criteria may include several categories such as:

Functional (related to business requirements)
Technology (related to the technologies/non-functional requirements).
Supplier capability and maturity
Delivery approach
Overall Cost.

Once the proposals have been received further short listing may take place to limit the formal evaluation to around 3–5 suppliers.

9.5 Evaluate Proposals and Select Supplier

The evaluation team will evaluate all received proposals using an evaluation spreadsheet (or similar mechanism), and the results of the evaluation yield a short list of around three suppliers. The short-listed suppliers are then invited to make a presentation to the evaluation team, and this allows the team to question each supplier in detail to gain a better understanding of the solution that they are offering, and to identify any risks with the supplier and their proposed solution.

Following the presentations and Q&A sessions the evaluation team will follow up with checks on reference sites for each supplier. The evaluation spread sheet is updated with all the information gained from the presentations, the reference site checks, and the risks associated with individual suppliers.

Finally, an evaluation report is prepared to give a summary of the evaluation, and this includes the recommendation of the preferred supplier. The project board then makes a decision to accept the recommendation; select an alternate supplier; or restart the procurement process.

9.6 Legal Agreement

The preferred supplier is informed on the outcome of the evaluation, and negotiations on a formal legal agreement commences. The agreement will need to be signed by both parties, and may (depending on the type of agreement) include (Fig. 9.1):

Legal Contract
Statement of Work
Implementation Plan
Training Plan
User Guides and Manuals
Customer Support to be provided
Service Level Agreement
Escrow Agreement
Warranty Period.

Fig. 9.1 Legal contract

The *statement of work* (SOW) is employed in bespoke software development, and it details the work to be carried out, the activities involved, the deliverables to be produced, the personnel involved and their roles and responsibilities.

A *service level agreement* (SLA) is an agreement between the customer and service provider, which specifies the service that the customer will receive as well as the response time to customer issues and problems. It will also detail the penalties should the service performance fall below the defined levels.

An *Escrow agreement* is an agreement made between two parties where an independent trusted third party acts as an intermediary between both parties. The intermediary receives money from one party and sends it to the other party when contractual obligations are satisfied. Under an Escrow agreement the trusted third party may also hold documents and source code.

9.7 Managing the Supplier

The activities involved in the management of the supplier are similar to the standard project management activities discussed in Chap 4 of [1]. The supplier may be based in a different physical location (possibly in another country), and so regular communication is essential for the duration of the contract. The project manager is responsible for managing the supplier, and will typically communicate with the supplier on a daily basis. The supplier will send regular status reports detailing progress made as well as any risks and issues. The activities involved in supplier management include:

Monitoring progress
Managing schedule, effort and budget
Managing risks and issues
Managing changes to the scope of the project
Obtaining weekly progress reports from the supplier
Managing project milestones
Managing quality
Reviewing the supplier's work
Performing audits of the supplier's work
Monitoring test results and correction of defects
Acceptance testing of the delivered software.

The project manager will maintain daily contact with the supplier, and will monitor progress, milestones, risks and issues. The risks associated with the supplier include the supplier delivering late or over budget, or delivering poor quality, and all risks need to be managed.

9.8 Acceptance of Software

Acceptance testing is carried out to ensure that the software developed by the supplier is fit for purpose. The supplied software may just be a part of the overall system, and it may need to be integrated with other software. The acceptance testing involves:

Preparation of acceptance test cases (this is the acceptance criteria)
Planning and scheduling of acceptance testing
Setting up the Test Environment
Execution of test cases (UAT testing) to verify acceptance criteria is satisfied
Test Reporting
Communication of defects to supplier
Correction of the defects by supplier
Re-testing and Acceptance of software.

The project manager will communicate any defects with the software to the supplier, and the supplier makes the required corrections and modifications to the software. Re-testing then takes place and once all acceptance tests have successfully passed the software is accepted.

9.9 Rollout and Customer Support

This activity is concerned with the rollout of the software at the customer site, and the handover to the support and maintenance team. It involves:

Deployment of the software at customer site
Provision of training to staff
Handover to the Support and Maintenance Team
On-going customer support
On-going maintenance.

9.10 Ethical Software Outsourcing

Software outsourcing is a way for a business to hire a third party subcontractor to develop all or part of a software development project, rather than carrying out the project in-house. It has become popular for western companies to outsource software developments to countries in Asia and Eastern Europe, with India now a major player for software outsourcing, and Poland and the Ukraine[1] have also become popular.

---

[1] This was prior to the Russian invasion of Ukraine in 2022.

9.10 Ethical Software Outsourcing

There are various motivations for outsourcing such as the desire to reduce the cost of software development, or it may be that the company may wish to focus on its core business and to outsource non-core activities, or it may be that the company lacks the expertise or capacity to implement the project internally. There are various models of outsourcing including where a company may partner with a third party supplier as a way to obtain extra IT resources for a company project, or it might outsource all or parts of the project to a third party supplier under the company's supervision, or it may outsource to a subcontractor that is given full responsibility for the work from the start to the end with minimal supervision.

The costs of outsourcing may be significantly cheaper than developing the software internally, but there are risks that it could be work out more expensive where there are delays or significant rework due to poor quality. There are risks of disruption of business activities depending on the political climate of the country where the subcontractor is based. Further, there may be risks of pandemics, natural disasters, or the subcontractor becoming bankrupt. It is essential that contractors are qualified for the work that they are to perform, and all associated risks must be managed.

The area of corporate social responsibility (CSR) has become topical in recent years, and companies have a responsibility to be good corporate citizens and to consider wider society in their actions and to consider their impact on the world. That is, corporations are expected to behave ethically, and to be environmentally conscious of their carbon footprint and on the sustainability of their business operations in the countries in which they are operating (even at the expense of profits).

There are several ethical issues with outsourcing such as the fact that outsourcing may lead to loss of jobs in the home country of the company, when it decides to outsource its software development to a cheaper country. It would seem reasonable to expect an ethical corporation to protect jobs in the countries where it is operating, and to avoid exploiting workers in low cost countries.

Ethical corporations have a responsibility to ensure that there are reasonable work practices in place at the subcontractor company, and that workers receive a fair salary, have reasonable terms of employment, and are not exploited by the subcontractor. Globalisation and the outsourcing of manufacturing operations led to many sweatshops in Asia, and there is the infamous case of worker exploitation at Foxconn (an Apple supplier of the *iPhone*) based in Shenzhen in China. Several Foxconn employees committed suicide due to their working conditions and their exploitation by the company, and this raised important ethical questions on the responsibilities of Apple for the health and welfare of the employees of one of its key suppliers. It is reasonable to expect a company as profitable as Apple to ensure that the employees of its suppliers have reasonable conditions of employment and that they are not exploited.

Advanced economies have many laws and regulations to protect the environment, as well as laws to protect the health and safety of employees in the workplace. However, the laws and regulations in Africa and Asia or wherever the subcontractor is based may not be stringent. An ethical corporate citizen has responsibilities to the environment, and it is not sufficient for the corporation to say that it is complying fully with the laws of every country it is operating, where those laws are not fit for purpose. That is, an ethical corporate citizen must do no harm (i.e., *primum non nocere*), and this is more than just complying with the laws of the country. It needs to ensure that no damage (e.g., pollution or environmental) arises as a result of its activities. Further, the corporation has ethical responsibilities for the health and safety of the subcontractor staff that are working on their projects.

There may be significant cultural differences between the country where the corporation is based and the country where the subcontractor is based, and there may potentially be very different ethical values in both countries. There may be problems with the political system in the country where the subcontractor is based, where an authoritarian government may maintain a strong control over the state and its citizens.

There may be problems with corruption, where bribes are paid to officials and others to get things done and to remove roadblocks. There may be unethical practices over price fixing, as well as cultural differences on the understanding of the importance and protection of proprietary information, intellectual property, and compliance to security and privacy standards. There may be weak environmental controls in the country where the subcontractor is based, and an ethical corporation will seek to ensure that the environment is protected. It is important to be explicit on expectations in the software outsourcing (in the legal contract) so that there is no room for misunderstanding on ethical values.

A corporation will generally wish to seek the cheapest offering, but it is also important for it to consider the ethical impacts of outsourcing. An ethical corporation will need to check the ethical behaviour of the subcontractor on a periodic basis including salary and working conditions, and one way to do this is to perform audits of suppliers. Audits provide visibility into the technical software development work being done to verify its compliance to standards and all appropriate laws and regulations, and special ethical audits could be conducted to provide insight into any work practices that could create ethical difficulties.

It is generally inappropriate to award a contract to a subcontractor just on price alone, and while price is an important criterion it is just one among many criteria, and ethical criteria should also be considered in the evaluation. It is best to build a stable relationship with suppliers, where there is a deep understanding of each supplier and any associated risks (including ethical risks), and to monitor and manage these risks.

9.11 Legal Breach of Contract

The legal agreement between the company and the subcontractor specifies the terms to be satisfied and the obligations on both parties for the duration of the contract. These include the deliverables to be produced, the timelines, the responsibilities of both parties, and the financial payments to be made at agreed milestones.

9.11 Legal Breach of Contract

A contract is legally binding on both parties with both having defined obligations, and should one party fail to deliver according to the terms of the agreement then they may be in breach of the contract.[2]

A *material breach* is where one party does not fulfil their obligations under the contract, or delivers a significantly different result from that defined in the contract. An *anticipatory breach* is where one party has indicated that they will not be fulfilling their obligations under the contract, and while an actual breach has not yet occurred there is an intention to be in breach of the contract. Both parties will generally discuss and attempt to resolve any such breaches, and it is generally easy to resolve *minor breaches*. However, if both parties are unable to resolve their dispute over a material breach in the contract, then one party may decide to sue the other party for being in breach of contract. However, legal disputes tend to be expensive and time consuming, and it is often more economical and in the best interest of both parties to resolve their dispute without the involvement of their lawyers.

However, if both parties are unable to find a reasonable resolution to their dispute the plaintiff may decide to bring the lawsuit to court claiming a material breach in the contract. The plaintiff will need to show that there was a legally binding contract between both parties, that the plaintiff fulfilled all of his obligations under the contract (unless there was a legitimate reason not to), that the defendant failed to honour the terms of the legal agreement, and that the defendant's actions led to loss being suffered by the plaintiff. That is, the breach of contract claim involves proving that:

Existence of contract
Plaintiff honoured contract
Defendant did not fulfil conditions of contract
Plaintiff suffered loss or damages.

The court will need to decide if there was a material breach of the contract, and will consider the arguments made by both the plaintiff and the defendant. The defence may argue that misunderstandings, misinterpretations and errors in the terms of the contract signed by both the plaintiff and defendant led to the breach of contract, and the judge will need to weigh up and consider all of the evidence and issue a judgment. The judgment is based on the facts of the case and the details of the contract, and it may be in favour of the defendant or the plaintiff depending on the circumstances of the case. For example, if the judge decides in favour of the plaintiff the remedy may be restitution and could potentially include:

Injunction to compel the defendant to provide restitution
Award of financial compensation for the breach of contract
Punitive damages (where appropriate) to punish the wrongdoer.

---

[2] It is also possible that two parties make a verbal contract that is legally enforceable.

Table 9.2 Possible breaches of contract

Breach	Description
Missing/late Deliverables	This is where the supplier has failed to deliver one or more deliverables, or where they have been delivered late
Deliverables not fit for purpose	This is where one or more deliverables do not satisfy the requirements, or they may fail to adhere to the defined standards or may be of poor quality or unusable
Missing Personnel	This is where the agreed human resources for the contract have not been provided
Unskilled Resources	This is where the resources provided lack the skills and experience to perform their roles effectively
Inadequate development environment	This is where the software engineering environment provided is not fit for purpose for the development and testing of the software
Intellectual Property not protected	This is where the intellectual property (e.g., patents and copyright) has not been properly protected
Proprietary information not protected	This is where the confidentiality of proprietary information provided to the subcontractor has not been protected
Quality Problems	This is where there are serious quality problems in testing or with the software produced, and where the software does not perform correctly under real world conditions
Inadequate Support (SLA)	This is where the support provided has been below the level agreed between the parties. It may be that the resolution of problems has not achieved the targets in the service level agreement
Bankrupt supplier	This is where the supplier has become bankrupt and is unable to fulfil their obligations

There are many possible breaches that could occur such as (Table 9.2).

9.12 Review Questions

1. What are the main activities in supplier selection and management?
2. What factors would lead an organization to seek a supplier rather than developing a software solution in-house?
3. What are the benefits of out-sourcing?
4. Describe how a supplier should be selected.
5. Describe how a supplier should be managed.
6. What is a service level agreement?

7. Describe the purpose of a statement of work?
8. What is an Escrow agreement?
9. What is ethical outsourcing?
10. What is a breach of contract?
11. What does a claim for breach of contract involve?

9.13 Summary

Outsourcing is a common business practice where a company contracts out business functions such as manufacturing, software development, and call centres to third party providers.

Supplier selection and management is concerned with the selection and management of a third-party software supplier. Many large projects often involve total or partial outsourcing of the software development, and it is therefore essential to select a supplier that is capable of delivering high-quality and reliable software on time and on budget.

This means that the process for the selection of the supplier needs to be rigorous, and that the capability of the supplier is clearly understood, as well as knowing any risks associated with the supplier. The selection is based on objective criteria, and the evaluation team will rate each supplier against the criteria, and recommend their preferred supplier.

Once the selection is finalised a legal agreement is drawn up which includes the terms and condition of the contract as well as a statement of work. The supplier then commences the defined work, and is appropriately managed for the duration of the contract.

The project manager communicates with the supplier on a daily basis during the contract, and manages issues and risks. The software is subject to acceptance testing before it is accepted from the supplier.

The legal agreement between the company and the subcontractor specifies the obligations on both parties for the duration of the contract. A contract is legally binding on both parties, and should one party fail to deliver according to the terms of the agreement then they may be in breach of the contract.

It is often inappropriate to award the contract to a subcontractor just on price alone, and ethical criteria should also be considered in the evaluation. An ethical corporation should act as a corporate citizen with responsibilities to the environment, and it is not sufficient for the corporation to say that it is complying fully with the laws of every country it is operating.

Reference

1. G. O' Regan, *Concise Guide to Software Engineering*, 2nd edn (Springer, Heidelberg, 2022)

Intellectual Property Law

10

Abstract

This chapter discusses intellectual property including patents, copyrights and trademarks. Intellectual property law deals with the rules that apply in protecting inventions, designs and artistic work, and in enforcing such rights. Patents protect innovative ideas and concepts, and give inventors exclusive rights to their invention for a specified period or time. A copyright applies to original writing, music, and other original intellectual and artistic expressions. It protects the expression of the idea and not the underlying idea itself. A trademark protects names or symbols that are used to identify goods or services, and their purpose is to avoid confusion and to help customers distinguish one brand from another.

Keywords

Copyright • Open source software • Fair use • Patents • Trademarks • Trade secret

10.1 Introduction

Intellectual property law deals with the rules that apply in protecting inventions, designs and artistic work, and in enforcing such rights. Intangible assets such as software designs or inventions may be protected in a similar way to the protection of private property, and an inventor is generally granted exclusive rights to the invention for a defined period of time. This gives the inventor the incentive to develop creative works that may benefit society, as it allows the owner of the invention to profit from their work without fear of misappropriation by others.

The main forms of intellectual property are patents, copyright and trademarks. Patents give inventors exclusive rights to their invention for a specified period (possibly up to 20 years), or the inventor may profit from the invention by transferring the rights to another party. A *patent* protects innovative ideas and concepts, and the invention itself must be novel and more than an obvious next step from existing technology. The patent needs to be filed at the Patent Office, and the patent gives the inventor protection against patent infringement in a specific country or region of the world.

A *copyright* applies to original writing, music, motion pictures and other original intellectual and artistic expressions. It does not protect the underlying idea as such, and what is protected is the expression of the idea. Copyrights are exclusive rights to making copies of the expression, where the ways of expressing ideas is copyrightable. Copyright law protects computer software source code from being copied by third parties without obtaining the required permission. The term *"fair use"* refers to the permitted limited use of copyrightable material without acquiring permission from the copyright owner.

A *trademark* protects names or symbols that are used to identify goods or services, and their purpose is to avoid confusion and to help customers to distinguish one brand from another.

A *trade secret* is information that provides competitive advantage over others, and it is of value only if it is kept secret. It applies in the computer sector where programs may use algorithms that are unknown to others.[1]

10.2 Patents

Patents are the strongest part of intellectual property and they provide much stronger protection than either copyrights or trademarks. They protect the implementation of innovative ideas and inventions for a period of time (often up to 20 years), but the time interval of protection is much shorter than with copyrights (a copyright is valid for 70 years after the death of the author) or trademarks (often for an indefinite time period). Patents are expensive to apply for and costly to obtain/defend, but they in effect give the inventor a monopoly (or exclusive rights) to exploit the invention during the lifetime of the patent, and all others need the permission of the patent holder should they wish to use the invention. At the end of the lifetime of the patent the invention becomes public domain with unrestrictive use (Fig. 10.1).

A patent may be for an innovative idea for a product or a process for making something in a new way. There are several types of patents such as *utility patents* that are patents for a novel, useful and functional invention. These are the most common type of patent, and are the most useful and profitable. There is a variation termed the *utility model* for small useful functional innovations. These patents are narrow in focus and so do not merit a utility patent as such, and they often refer to something small in the manufacturing process. These tend to be inexpensive and fast to obtain, and they are often called petty or minor patents. *Design patents*

---

[1] It is a grey area on whether it is legal to use reverse engineering to try to discover the trade secret.

Fig. 10.1 Patent for an invention

protect the beauty or design of a manufactured product, and so it is essentially like the copyright of the design of the manufactured product. It protects the appearance of the product but not its functionality.

Another words, a utility patent protects the idea or function (the way it works) of something useful such as the invention of a product or a process, whereas a design patent protects the expression of the idea such as the appearance of the product. For example, the *iPhone* has over 2000 utility patents protecting how various parts of the phone work, but it only has a small number of design patents that protect its appearance. Often utility patents apply to sub-elements of the product rather than to the product itself, and so instead many functions of the product are patented, as the product itself is too big to patent.

The benefits of patents to society are that they encourage investment and the development of useful products, and foster a culture of innovation in the state, as the inventor and wider society benefits from the exploitation of the invention. Trade secrets were a common way to protect inventions prior to the widespread use of patents, but their disadvantage is that other companies were unaware of the latest developments.

The granting of exclusive rights to the inventor for a temporary period of time (the lifetime of the patent) fostered the publication of inventions, and encouraged their commercial exploitation as the inventor had exclusive (or monopoly rights) for a period of time. This is especially important in situations where there are large costs and major risks involved in bringing a product to market. Further, the invention or technology could be licensed (for a fee) to other companies during the lifetime of the patent. For example, there are many patents on mobile phones, and companies may license their patents to the manufacturers that make the phones.

The disadvantage of patents is that they may impact competition as they act as a barrier to market entry. A market participant may be reluctant to enter the market due to the high charges incurred for each licensed patent, and existing participants may be discouraged with these charges and depart the market. It may also lead to higher charges for consumers, as there are significant legal costs involved in both lodging and defending patents, and these charges are passed on to consumers. This means that the cost of the product is higher than it would otherwise be due to the legal overhead associated with patents.

Patents are difficult and expensive to apply for, and the patent application is time consuming for both companies and the government patent office. Further, lawsuits are expensive in both taking a lawsuit against another party for patent infringement, and in defending a lawsuit taken by another company that is alleging patent violation. And so, the question is sometimes asked as to whether the benefits of patents are worth the costs, with some computer scientists such as Richard Stallman arguing against intellectual property law such as patents (see Chap. 52 of [1]). The process for obtaining a utility patent involves several steps as described in Table 10.1.

The patent application must show that the invention is *novel*[2] (i.e., the invention is new and the inventor is the first to discover it). Next, it must be more than an obvious next step from existing technology (i.e., the invention is *not obvious* and is a significant advance over the existing *prior art*[3]). Further, the invention must have *utility*, and so it must be useful and be of practical benefit to the public and society. It is important that there is a good business case for the patent, and a desirable characteristic is when competitors are unable to bypass the patent. Another word, the basic requirements to make an application for a patent for an invention are that it is:

- Novel
- Utility
- Not obvious
- Has a good business case
- Difficult for competitors to bypass invention.

The process for obtaining a design patent is easy and quick to get as it is just protecting appearance, and all that it required is to show that it is new or original, that the person who is filing the patent is its creator and has not copied it from someone else, and that it is ornamental. The process for obtaining a utility model

---

[2] The invention must not have been publicly disclosed to the public (e.g., described in a publication or presented at a conference).

[3] Prior art refers to the existing state of knowledge of a field.

Table 10.1 Process for obtaining a patent

Step	Description
Obtaining a Provisional Patent	This starts the clock on the patent application, and although this step may be skipped it is best not to, as it gives all the rights to the patent applicant. The date of first application is extremely relevant in patent law, as priority is given to the party that is first to file. Another words, if another party files before this date then they have priority and all rights to the invention. The provisional patent is easy to obtain and does not need to be lodged by a patent attorney, and it gives the right to the inventor to say: *"patent pending"*
Formal Application	The formal patent application may be filed up to one year from the provisional application, and it requires a lot of technical detail, and is expensive (up to $100 k) and time consuming. The application must provide evidence and detail to show that the invention is novel, useful, and more than an obvious next step from existing technology. The majority of patents filed will be rejected
Review and Appeal	The third step involves working through the patent review and appeal process (average 2.5 years) at the patent office. The first step for the patent attorney is to discuss the rejected patent with the patent officer/internal review with supervisor/patent board to understand if there is something missing, and deal with any questions If this fails then the next step is to take legal action against the patent office, with the goal of getting the judge to overrule its decision. If the judge upholds the decision then the next step is to appeal the judgment to a higher court until all legal options are exhausted There is then a final ruing on whether the patent should be granted or rejected
Defending Patent	Once a patent is granted there may be a need to go to court to defend it (e.g., another party might dispute the validity of the patent and argue that it should be revoked) anytime during its lifetime, or to take action against a party for patent infringement (e.g., a party may use a patent without permission) Patents that have been granted may be overturned on appeal

patent is short and quick, but the protection granted is not as strong as a utility patent.

A patent is valid in only one jurisdiction, where a jurisdiction may be part of a country (e.g., Macau is part of China but has its own system), or more than one country (e.g., the jurisdiction of the EU covers several countries). This differs from the protections granted under copyright law where a copyright is valid globally. There are some international treaties that give patent protection rights across country boundaries, and there are also some international patent applications that allow for the application of patents in more than one country at the same time, including a centralised patent review, and the issuing of multiple patents.

However, the general rule is one patent per country, and the patent can only be enforced in the country or jurisdiction where it was issued. Another words, a patent issued in Malaysia is not valid in the United States and vice versa, and so there is a need for the company or individual to apply for a patent in every country that is important. Further, if the patent is issued to a company or individual in Malaysia then it can be only enforced with a lawsuit in Malaysia, and the courts will not consider patents issued from other countries.

10.2.1 Filing a Patent

The motivation of a company in seeking a patent is to protect its commercial innovations from its competitors, and the patent gives the inventor protection against infringement by others, as well as exclusive rights to exploit the invention for a period of time. The patent needs to be precisely described and the invention must be novel, which means that it is something new that has not been created by another, and it must not have been publicly disclosed to the public (e.g., described in a publication or presented at a conference). There is a grace period of one year in most countries where a patent application[4] may be made following a public disclosure, but if the patent application is made (even one day) after this then it is invalid and the invention is considered to be in the public domain.

The invention must have utility and so it must be useful to the public, and so the formal application must state the way in which the product or process is useful. That is, there is at least one useful application stated and there may be multiple applications.

The invention must not be obvious, and so it is more than an obvious next step from the existing technology. That is, it must be more than a transpose of existing technology, and should be a significant advance over the existing prior art. The application should include a description of the current state of the art, as well as a technical description of the invention. It needs to highlight the applications and advantages of invention, and relevant drawings should be included.

Most of the effort in a patent application goes into describing how the invention is not obvious, and so there is a lot of detail on the existing prior art, and how the new invention is a significant advance on existing inventions. All relevant, pertinent and analogous information must be included in the application, and there is a test applied by the patent office to determine the extent to which the invention is not obvious:

- Survey existing prior art
- Examine differences invention/prior art

---

[4] This may be a provisional application: i.e., it does not need to be a formal application.

- Describe level of skills required to measure whether obvious or not.

There must be a good business case for the patent given the costs involved in filing and defending it. This means that the patent needs to be written in such a way so that it becomes hard for competitors to bypass or get around the invention. Otherwise competitors will be able avoid the payment of a license fee for its use, and this will reduce the royalties earned from the patent, and limit its commercial return to the company. The best patents are often the smallest patents, and it is often best to have a lot of small patents as close to the function or element of the product rather than having one large broad patent.

A business decision needs to be made (based on costs/benefits) once the patent has been described at the right level of detail on whether to go ahead with the formal application to the patent office or not.[5] It may be that there is a lack of a business case (insufficient return on investment) to file an application at the patent office, and that the most appropriate way to proceed is to place the invention in the public domain. The best way to do this is with a defensive publication that describes the invention and places it in the public domain for all to use, and thereby preventing a competitor from lodging a patent application.

The application for the patent will then be filed at the Patent Office, and the patent officer will examine the patent application and review and search for prior art to determine the extent to which the invention is novel, has utility, and is not an obvious next step with the existing technology. The patent office grants (or rejects) the patent, and in the case of a rejection discussions take place between the applicant and the patent officer as well as an internal patent review with the goal of overturning the decision. Legal action may be taken against the patent office in the courts (up to the supreme court) with the goal of overturning the patent office decision. Finally, in the event of a patent finally being granted the inventor may earn a royalty fee from the invention for a defined period (based upon its use), and the inventor may need to defend the patent.

10.2.2 Patent Litigation

Patent litigation is where the patent owner takes a lawsuit against another party for infringing the exclusive rights that the patent holder has with respect to the invention. The defendant may be selling a technology that uses the patented invention, and the plaintiff will need to show that:

- It has a patent for the invention
- The patent has not expired

---

[5] The decision may be to put the invention in the public domain with a defensive publication thereby preventing competitors from filing a patent for the invention. The costs of filing a formal patent application may be up to $100 k.

- The patent is valid in the particular country
- The defendant used the invention
- The defendant did not have a license to use the invention
- The defendant's actions led to loss to the plaintiff.

Often, the parties will reach a settle (e.g., licensing) rather than going through the expense of the legal process.

10.2.3 Case Study—Patent Ruling on First Computer

The Atanasoff-Berry Computer (ABC) was one of the earliest electronic digital computers,[6] and it was approximately the size of a large desk and consisted of approximately 270 vacuum tubes. Two hundred and ten tubes controlled the arithmetic unit; 30 tubes controlled the card reader and cardpunch; and the remaining tubes helped maintain charges in the condensers. It employed rotating drum memory, with each of the two drum memory units able to hold thirty 50-bit numbers (Fig. 10.2).

The ABC was a digital machine that was designed for a specific purpose (i.e., solving linear equations) rather than as a general-purpose computer. However, it was slow and required constant operator monitoring [2].

It used binary mathematics and Boolean Logic to solve simultaneous linear equations. It employed over 270 vacuum tubes for digital computation, but it had no central processing unit (CPU), and it was not programmable. It weighed over 300 kg and it used 1.6 km of wiring. It used 50-bit numbers, and it could perform 30 additions or subtractions per second. The memory and arithmetic units could operate and store 60 such numbers at a time (60 * 50 = 3000 bits). The arithmetic logic unit was fully electronic, and it was implemented with vacuum tubes.

The input was in decimal format with standard IBM 80 column punch cards, and the output was in decimal format via a front panel display. A paper card reader was used as an intermediate storage device to store the results of operations too large to be handled entirely within electronic memory. The ABC pioneered important elements in modern computing including:

- Binary arithmetic and Boolean logic
- All calculations were performed using electronics rather than mechanical switches.
- Computation and memory were separated

---

[6] The ABC was ruled to be the first electronic digital computer in the Sperry Rand versus Honeywell patent case in 1973. However, Zuse's Z3 computer preceded it (as it was completed in 1941 whereas the ABC became operational in 1942).

10.2 Patents

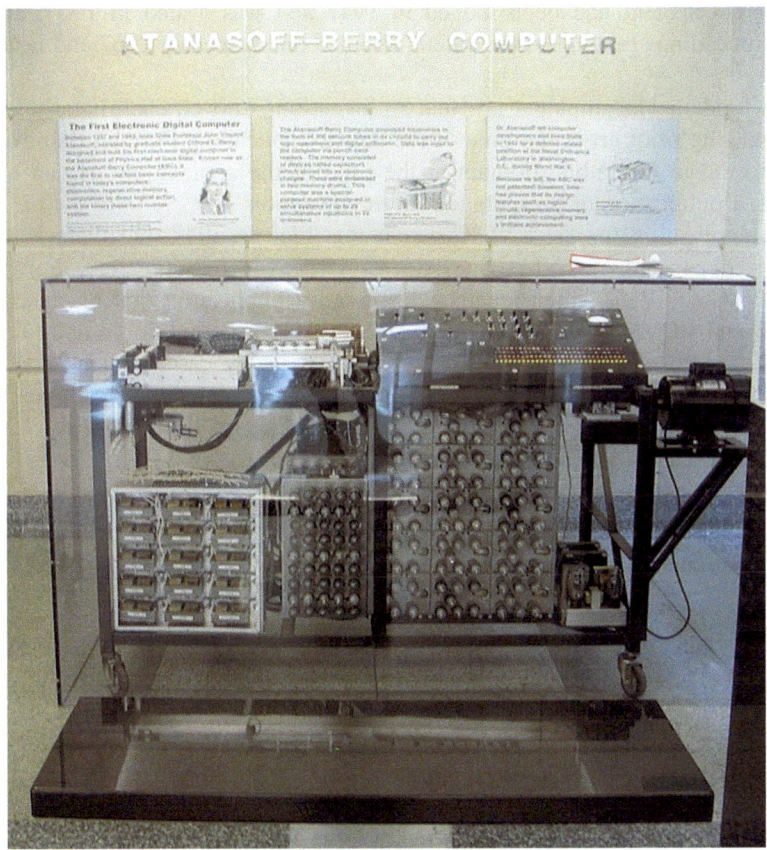

Fig. 10.2 Replica of ABC computer: creative commons

The ABC was tested and operational by 1942, and its historical significance is that it demonstrated the feasibility of electronic computing. Mauchly and Eckert of the Moore School at the University of Pennsylvania later used several of its concepts in the development of the ENIAC computer.

The ABC computer was ruled to be the first electronic digital computer in the 1973 *Honeywell versus. Sperry Rand* patent court case in the United States. The court case arose from a patent dispute between Sperry and Honeywell,[7] where Sperry was accusing Honeywell of patent infringement, and demanding compensation and royalties. Honeywell counter sued and charged Sperry with monopoly and fraud, and demanded that the Sperry patent on ENIAC be declared invalid.

---

[7] Sperry (later called Unisys) was one of the earliest computer companies and it was the successor to EMCC that was founded by Presper Eckert and John Mauchly (the designers of ENIAC and EDVAC in the mid-1940s). Honeywell Information Systems was founded in the mid-1950s.

The ENIAC patent had been lodged in 1947 and was issued in 1964, and the legal proceedings relating to the patent dispute commenced in 1967 and lasted for 6 years.

It is fundamental in patent law that an invention is novel, useful and that there is no existing prior art at the time of the patent application. Further, the invention must not be in the public domain at the time of the application, as can happen through a publication or presentation on the invention.

The application for the ENIAC patent was filed in 1947, but there had been a public disclosure of ENIAC in 1946, as well as Von Neumann's draft report on EDVAC in 1945, which legally constituted a publication that disclosed both ENIAC and EDVAC, and in effect placed ENIAC in the public domain. Further, John Atanasoff was called as an expert witness in the case, and the court also ruled that Eckert and Mauchly did not invent the first electronic computer, since the ABC existed as *prior art* at the time of their patent application for ENIAC. This meant that the Mauchly and Eckert patent application for ENIAC was invalid, and John Atanasoff was named as the inventor of the first digital computer.

Mauchly had visited Atanasoff on several occasions prior to the development of ENIAC, and they had discussed the implementation of the ABC computer. Mauchly subsequently designed the ENIAC, EDVAC and UNIVAC computers. The court ruled that the ABC was the first digital computer, and that the inventors of ENIAC had derived the subject matter of the electronic digital computer from Atanasoff.

10.3 Copyright

Copyrights apply to original writing and to original intellectual and artistic expressions, and it protects the expression of the idea rather than facts or the idea itself. Copyright law protects literary, musical and artistic works such as poetry, songs, books, painting, dance, movies, music, information in news media and computer software. It provides exclusive rights to making copies of the expression (subject to copyright law and fair use), where the ways of expressing ideas is copyrightable.

One of the earliest disputes in copyright law occurred during early Irish Christianity in the late sixth century A.D., where there is a story of a dispute between St. Columba and St. Finnian over the right to copy part of the bible. St. Jerome had created a Latin copy of the bible called the Vulgate, and St. Jerome's Psalter refers to the Book of Psalms in the bible. St. Columba had borrowed the Psalter from St. Finnian, and made a copy that he called the *Cathach*. St. Finnian disputed St. Columba's right to make a copy, and he claimed ownership of the copy. King Diarmuid Mac Cerbhaill intervened to resolve the dispute, and he ruled that "*To every cow belongs her calf, therefore to every book belongs its copy*", and so established copyright law in early Christian Ireland (Fig. 10.3).

A copyright gives the copyright owner rights to exclude others from using or copying the finished work, and most copyrights are valid for the creator's lifetime plus 70 years (the exact period depends on the jurisdiction as copyright laws vary

10.3 Copyright

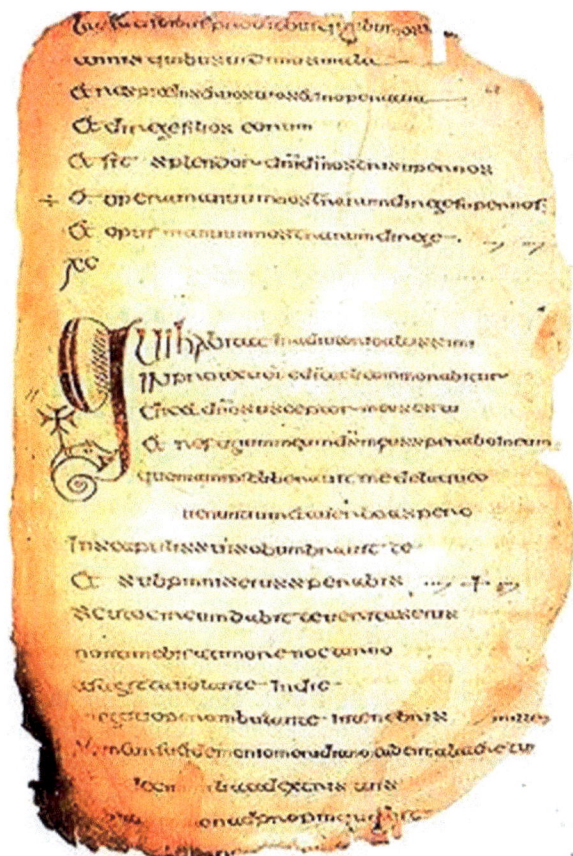

Fig. 10.3 St. Colomba's Cathach

between countries). The original reason why copyright developed was to motivate artists to produce more artistic work thereby encouraging creative art, and so artists are rewarded for creating more music, art, and so on. Over time, the period of time for protection has increased in a major way, and today the purpose of copyright is more about protecting the artistic creations of business and large corporations.

The creator of an original work may obtain a copyright, and the work needs to be recorded (e.g., on paper, art, laptops, and mobile phones). Copyright is automatic in most countries although a small number of countries require registration. Registration gives extra benefits as often it allows the copyright owner to sue for a larger amount, and so it may be useful for the creator of the work to get registered. The work must be:

- Original
- Recorded
- Registration is useful
- Mainly corporate owned.

Copyrights do not protect ideas or concepts and these require the protection of a patent, whereas copyright law protects the expression of the idea and not the idea itself. Sometimes in a copyright dispute one party may be alleging that the expression protected in the copyright was violated, with the other party arguing that it was the idea that was used to create a derivative work and that no copyright violation took place. Names and common phrases may not be protected, and similarly with facts and data. Methods of operation or equipment maintenance instructions are excluded under copyright law, as are most manufactured goods (exceptions include books and DVDs). Useful things are not protected by copyright as patents protect most of these, but there are some exceptions such as dictionaries and software where expression is protected.

The copyright holder has several rights including the right to *prevent others from making copies* of the work, as well as the rights to *stop others from making a derivative work*, i.e., something that is based upon the work. The copyright owners have the *right to distribute the work*, and to display the work anywhere and anytime they like. The owners of a copy have limited rights and may display the copy in one location only. The original *copyright owner has performance rights*, and the *right to exclude others from performing the work*. The performance right is limited in the case of music, where others have the right to perform the music in public in return for a royalty payment to the original author of the song.

There are several limitations of copyright and the most important of these is "*fair use*" (or right to copy), which refers to the permitted limited use of copyrighted material without obtaining permission from the copyright owner. Fair use is not a right as such: rather, it is more of a defence that the defendant makes to the judge, and it is the rationale for why the use of the copyrighted material is viewed as fair by the defendant. Fair use is a little subjective and there are several factors that the judge will need to consider in coming to an informed and balanced view on whether it is reasonable to apply fair use to the particular case. These factors include:

- Purpose of use
- Amount used
- Nature of work
- Commercial impact.

The purpose of use is the reason why the material has been used, and it is easier to justify fair use for non-profit purposes such as educational use (especially in face to face education rather than on-line education). Other areas that are used to justify fair use include literature, criticism, parody and news reporting. The amount or proportion of material used is important, as it is easier to justify the use of a small portion of the work, or a small proportion of the amount used to the whole of the copyrighted work. There may be complications if the amount used is the core part of the copyrighted material.

The nature of the work is important as some types are better protected than others, and so if the work is mainly facts and data it may not be so well protected,

Table 10.2 Exceptions to copyright

Exception	Description
First sale doctrine	The *first sale doctrine* allows an individual who purchased a copyrighted work the right to display or sell that particular copy. For example, the purchaser of a book may read it, record some notes in the book, and re-sell the book. This also applies to borrowing books from a library, where the borrower becomes the temporary owner of the book for the duration of the loan until it is returned to the library. Similarly, the purchaser of a software package may have the right to modify it for home use. However, it has become a common practice to distribute commercial software by licensing it rather than selling it, so that the user is unable to copy it and use the first sale doctrine as a defence
Performance rights limit	The copyright holder has exclusive rights to perform their work, but *performance rights limits* means that these rights do not apply if the performance is for non-commercial purposes such as for charity, religious purposes, and education
Consumer good will	*Consumer goodwill* restricts the ability of a company to act even where it has the legal right to do so, as so acting may damage consumer good will
Expired copyright	*Expired copyright* refers to a copyright that has expired (e.g., it is over 70 years since the author's death or the author may have placed the work in the public domain). Another words, an expired copyright no longer has any legal effect and the works of the author are in the public domain

whereas if it is a good story about a wizard called "Harry Potter" it will be well protected. The commercial impact involves determining the financial costs and the effect on the market or the commercial value from using the material, and if the costs are minimal it is easier to justify fair use. The defendant bears the burden of proving fair use in any litigation on copyright infringement. There are other exceptions to copyright including (Table 10.2):

A *copyright infringement* is where the rights of the copyright owner have been violated, and where there may be grounds to sue another party for infringing these rights. An indirect infringement is where a person indirectly and illegally uses copyrighted work (without being aware of it). For example, George Harrison indirectly used copyrighted work with his 1970 song "*My Sweet Lord*", where he unconsciously plagiarised the song "He's so fine" that was written by Ronnie Mack and recorded by the Chiffons in 1963. A direct infringement is where a person distributes, displays or performs copyrighted work, or prepares a derivative work without permission. A contributory infringement is where a person has contributed in some material way to the copyright infringement. The steps open to the copyright owner in a legal infringement case include:

- Show that it is the copyright owner
- Show copyright is valid
- Present evidence of infringement.

The copyright owner will need to convince the judge that a copyright infringement has occurred, and the plaintiff must first show that it is the copyright owner of the original work and that the copyright is still valid. Often, there may be no direct evidence of infringement, and so the judge often considers circumstantial evidence, as this may assist in determining on the balance of probability that it is more likely than not that an infringement took place. The judge will consider whether copying took place, and if so, whether that copying was legal (e.g., fair use, ideas, public domain) or not. Further, the more the works differ, the higher the standard of proof that is required to show infringement.

The damages claimed could include the economic damage such as the loss of sales and profit, or the plaintiff may be looking for statutory damages which are a high fixed amount to compensate for injury or loss as defined by law/statue. In the case of a blatant violation of copyright (i.e., an intentional or wilful infringement) the damages sought could include personal liability, including major costs for the individual or even that of a criminal offence.

It is permitted to create a derivative work if permission has been obtained from the author of the original work, and the agreement will often include a financial settlement for licensing to create the derivative work (e.g., creating a movie based upon a book will generally include a payment to the author). In many countries the translation of a book creates a derivative work, and so it requires the permission of the copyright holder, and similarly for audio books (China is an exception as it considers translation to be a transformative fair use). The recording of sheet music is a derivative work and needs permission, as are the arrangements of a work.

A parody is a derivative work that comments and makes fun of another work, and it is generally legal in that a successful defence of fair use is generally easy to make should a legal case be taken against the offender. However, such a defence may sometimes fail if the derivative work is rude or offensive, or due to differences in humour or culture. It is generally not required to seek permission to create a parody of a work.

10.3.1 Copyright for Software

Computer software is protected by copyright with software code protected automatically without copyright registration in most countries, and software copyright law is part of global copyright law. A copyright protects expression (i.e., the way in which something is said), and a registered copyright is inexpensive and easy to get, although court cases for copyright violation are expensive. Software patents protect function (not the expression of the function) and provide stronger protection, but they are costly and time consuming as well as being difficult to obtain.

The same function may be expressed in multiple ways by different programs, but if the function is protected (as in a software patent) then nobody else can do that particular thing without permission or licensing from the patent holder. That is, copyright law protects one particular expression of the function, and as a different program could implement the same function in a different way, the function itself is not protected. Therefore, if it is important to protect the function itself rather than just one expression of the function, then it is necessary to protect it with a patent, as copyright law protects just one expression of the function.

Computer software source code was granted protection by copyright law from the mid-1970s, which means that the reproduction of the computer software created by software developers and software companies is protected. This protection also includes that of the deliverables created as part of the software development process (e.g., the specification, design, code, testing and other artefacts). The work of an employee automatically belongs to the company (i.e., the company owns the creative work of its employees), and if an employee changes jobs then the intellectual creations of the employee's previous employment belong to the previous employer, and should not be disclosed to the new employer. The employee may have signed a non-disclosure agreement with the previous employer, and so the employee must respect any confidential information from previous employment.

Further, if a software contractor produces software code for a company then this generally belongs to the company (there is usually a signed legal agreement to that effect). All such work is protected by copyright, which means that if a software contractor is implementing the same function in another company, and uses the same software code that he/she previously developed for a former client, then the contractor and the new company would be in breach of copyright law, unless permission has been obtained for its use.

The copyright grants the author the right to exclude others from making copies, and the owners of the copies have the right to make additional copies (for archival purposes) without the authorisation of the copyright owner. However, as the owners of copies have the right to sell their copies, the software sector has moved towards licensing their software rather than selling it.

There is some software code that is freely available, and this includes software created by the free software movement (which began in the mid-1980s), the open-source initiative (which began in the late 1990s as a movement that wished to highlight the benefits of freely available source code), or software that is in the public domain and that is therefore not subject to copyright. Open-source software (OSS) is software that is freely available under an open source license to study, change and distribute to anyone for any purpose.

The 1998 US Digital Millennium Copyright Act (DMCA) is an extension of copyright law for the United States, and one of its motivations was a response to the rise of e-commerce in the late 1990s. Lawmakers wished to protect and support the digital economy, and to enhance protections for copyright law in the United States. There are three main areas of protection in the act:

- Safe harbour
- Digital Rights Management (DRM)
- Prohibit Copy Protection Circumvention.

DMCA provides *safe harbour* protections to Internet Service businesses in the US, by adding extra provisions to make fair use far wider for these businesses. It is designed to protect online service providers (OSP) from copyright claims that could arise from the conduct of their end users, which copyright holders claim is infringing their copyright. The law protects online service providers from being prosecuted for copyright infringement (assuming that the provider was not part of the infringement), and the main beneficiaries of the act were companies such as Google, Facebook and other large platform players. The EU has rejected the safe harbour provisions in DMCA.

DMCA essentially allowed large platforms to force content providers to give up their content for free if they wanted it to be searchable on their platforms. This resulted in the platforms rather than the content providers benefiting most from the commercial exploitation of their content. That is, there is no way that a small content provider such as a writer or musician could sue an Internet giant such as Google or Facebook for copyright infringement, whereas these Internet giants reaped all of the benefits by targeting advertisers as part of the search for content[8] [3].

Digital Rights Management (DRM) refers to using digital coding to enact rights like fair trial version or limited version or restrictions on use, and it essentially involves the use of digital code to assign digital assess rights to users. That is, users are stopped from doing certain things using digital code, and DMCA is a legal tool to make DRM stronger in its protections of these rights, and it makes violation of these rights a criminal offence.

Finally, the *prohibition of copy protection circumvention* makes it a criminal offence to break encryption algorithms to gain access, and does not provide a fair use exception. DMCA remains unique to the United States, but any individual or company around the world that breaks encryption algorithms could potentially be sued and even jailed in the United States. That is, DMCA poses risks to international companies, and while copyright violations are mainly civil lawsuits DMCA creates criminal liability for such violations.

10.3.2 Case Study—Copyright Dispute Apple Versus Microsoft

Apple Computers took a copyright infringement lawsuit against Microsoft in 1988, with the goal of preventing Microsoft from using its GUI elements in Microsoft's Windows operating system. The legal arguments lasted for five years, and the final ruling in 1993 was in favour of Microsoft.

---

[8] Another example of the digitalisation of content that benefited the Internet platforms rather than the content providers is the Google book project where Google intended that all the books in the world would be presented free on its platform. The book publishers and Google reached a financial settlement that gave the publishers a tiny financial payment compared to the revenue that Google gained from advertisers in becoming a massive digital bookstore where users could search for content. There was no benefit to authors and the benefits to the publishers were minimal. The publishers did not really understand Google's business model at the time, where Google's approach to the commodification of content led to the transfer of wealth from the content creators to the Google platform.

10.3 Copyright

Apple had claimed that the look and feel of the Macintosh operating system was protected by copyright, and this included 189 GUI elements. However, the judge found that 179 of these had already been licensed to Microsoft (as part of the Windows 1.0 licensing agreement), and that most of the 10 other GUI elements were not copyrightable.

This legal case generated a lot of interest as some observers considered Apple to be the villain of the piece, as they were using legal means to dominate the GUI market, and to restrict the use of an idea that was of benefit to the wider community. Others considered Microsoft to be the villain, with their theft of Apple's work, and their argument was that if Microsoft succeeded a precedent would be set in allowing larger companies to steal the core concepts of any software developer's work.

The court's judgment seemed to invalidate the copyrighting of the *"look and feel"* of an application. However, the judgment was based more on contract law rather than copyright law, as Microsoft and Apple had previously entered into a contract with respect to licensing of Apple's icons on the earlier Windows 1.0 operating system. Apple had not acquired a software patent to protect the intellectual idea of the look and feel of its Macintosh operating system, and it was actually Xerox PARC[9] (see Chap 35 of [4]) rather than Apple that had first developed the graphical user interface (GUI). There is a famous quote attributed to Bill Gates on the Apple / Microsoft legal case, where Gates is reported to said that there's more than one way of looking at the dispute in relation to the use of the Apple GUI and the subsequent Microsoft dispute. Gates compared their actions to them both having a rich neighbour called "Xerox", and that what happened was analogous to Gates breaking into the house to steal the TV and finding out that it was already stolen by Jobs.

> Well, Steve [Jobs]... I think it's more like we both had this rich neighbour named Xerox and I broke into his house to steal the TV set and found out that you had already stolen it.

---

[9] Xerox PARC's inventions have had a significant influence on developments in the computing field. However, Xerox has been criticized for failing to properly exploit its inventions, as other companies have reaped the benefits of its research. For example, the Xerox 8010 Star, its commercialisation of the Xerox Alto personal computer, was released in 1981 but it was not very successful due mainly to its high price of $16,000. Instead, it was Apple that reaped the benefits of PARC's research in personal computing, when it introduced the Apple Macintosh. This machine was a revolution in the computing field with its bitmap display and mouse driven graphical user interface, and these were copied from the existing Xerox Alto. The Apple Macintosh was an immediate success following its release in 1984.

10.3.3 Software Licenses

A software license is a legal agreement between the copyright owner and the licensee, which governs the use or distribution of software to the user (licensee). Computer software code is protected under copyright law in most countries, and a typical software license grants the user permission to make one or more copies of the software, where the copyright owner retains exclusive rights to the software under copyright law.

The two most common categories of software licenses that may be granted under copyright law are those for *proprietary software*, and those for *free open source software* (FOSS). The rights granted to the licensee are quite different for each of these categories of licenses, where the user has the right to copy, modify and distribute (under the same license) software that has been supplied under an open-source license, whereas proprietary software typically does not grant these rights to the user.

The *licensing of proprietary software* typically gives the owner of a copy of the software the right to use it (including the rights to make copies for archival purposes). The software may be accompanied with an end-user license agreement (EULA) that may place further restrictions on the rights of the user. There may be restrictions on the ownership of the copies made, and on the number of installations allowed under the term of the distribution. The ownership of the copy of the software often remains with the copyright owner, and the end user must accept the license agreement to use the software.

The most common licensing model is per single user, and the customer may purchase a certain number of licenses over a fixed period. Another model employed is the license per server model (for a site license), or a license per dongle model, which allows the owner of the dongle use the software on any computer. A license may be perpetual (it lasts forever), or it may be for a fixed period (typically one year).

The software license often includes maintenance of the software for a period (typically one year), and the maintenance agreement generally includes updates to fixes to the software during that time, and it may also cover a limited amount of technical support. The two parties may sign a service level agreement (SLA), which stipulates the service that will be provided by the service provider. This will generally include timelines for the resolution of serious problems, as well as financial penalties that will be applicable where the customer service performance does not meet the levels defined in the SLA.

Free and open-source licenses are often divided into two categories depending on the rights to be granted in distribution of the modified software. The first category aims to give users unlimited freedom to use, study and modify the software, and if the user adheres to the terms of an open source license such as GNU or General Public License (GPL), the freedom to distribute the software and any changes made to it. The second category of open source licenses give the user permission to use, study and modify the software, but not the right to distribute it freely under an open source license (it could be distributed as part of a proprietary software license).

10.3.4 Stallman and Free Software

Richard Stallman is an American computer scientist who is famous for being the *prophet of the Free Software Movement*. He is president of the *Free Software Foundation*, and has played a key role promoting the rights and freedom of software end users to use, copy, and modify software. The existing intellectual property rights for software are quite stringent, and Stallman has sought ways to maximize freedom for software end users (Fig. 10.4).

Stallman started the non-profit organization, the Free Software Foundation (FSF), in 1985, and there are thousands of volunteer programmers involved. They develop free software as part of the free software movement. He is the non-salaried president of FSF, and the meaning of the term *"free software"* in defined in the GNU manifesto. He lists four key freedoms essential to software development [5], and a program is termed *"free"* if it satisfies these properties. These are:

1. Freedom to run the program for any purpose
2. Freedom to access, study and to improve the code, and to modify it to suit your needs.
3. Freedom to make copies of the program and to redistribute them to others
4. Freedom to distribute copies of the modified program so that others can benefit from your improvements.

The GNU project uses software that is free for users to copy, edit, and distribute. *It is free in the sense that users can change the software to fit individual needs, and so not in the sense of "free beer"*. Stallman has written many essays on software freedom, and he is a key campaigner for the free software movement. The FSF has provided a legal framework for the free software movement, to protect the modification and distribution rights of free software. Stallman introduced the important concept of *"copyleft"*, which is a form of *licensing of free software*. It makes a program or product free, and requires that all modified or extended versions of the program are also free.

Fig. 10.4 Richard Stallman at Pittsburgh University

Stallman has argued against intellectual property such as patent law and copyright law. He has argued against patenting software ideas, stating that a patent is an absolute monopoly on the use of an idea. He states that while twenty years may not seem like a long period of time, that in the software field it is essentially a generation, due to the pace at which technology changes in the world we live in. Further, *patents act a barrier to competition and lead to monopolies.* They make it difficult for new companies to enter a market place, due to the restrictions and costs associated with the licensing of patents.

Stallman argues that copyright law places Draconian restrictions on the general public, and takes away freedoms that they would otherwise have. They protect the copyright owner businesses, and he suggests that the digital era requires us to consider alternative approaches.

10.4 Trademarks

Trademarks protect names or symbols that are used to identify goods or services, and help customers to distinguish one brand from another (e.g., "IBM" and "Microsoft" are well known brands that are protected by trademarks). Trademarks were originally employed to protect craftsmen, where the craftsman placed a physical mark on the product to signify that it was his work. This helped to avoid any confusion with other craftsmen who made a similar product. Trademarks play an important role in protecting business innovations, and trademark rights come from their actual use. The brand of a company is valuable (it forms part of the intangible assets on the company balance sheet), and trademarks play an important role in protecting the brand (or brand equity), as well as protecting the identity of the company.

The financial value of the intangible assets of the company's trademarks often exceeds that of the value of the value of the company's other intellectual property such as patents and copyrights combined. That is, trademarks are a large component of the value of large corporations.

Trademarks protect a company's logos (images associated with a brand), its names, slogans, its reputation and its brand. It is also possible to trademark a sound such as the sound that a mobile phone makes on ringing, or the sound that an operating system makes on start up. It is also possible to trademark the design of a car. There are some restrictions on what may be trademarked, and it is not possible to trademark something that is too generic, too descriptive or too functional.

A trademark protects the identity of a company and its reputation, and the value of a brand varies between companies. For example, the brand of "Apple" is much more valuable than the brand of a company such as "Lenova", and this allows Apple to charge more for its products than its competitors, as its customers view them as superior to competing products. This is due to the excellence of the design of Apple products and the value of the Apple brand, which enables Apple to have higher margins and to be more profitable than its competitors.

That is, companies with a strong brand generally get a premium price for their products, and are often more profitable than their competitors. This is since consumers are generally willing to pay the extra price due to the quality of the product, and the reputation and brand of the company. It takes a small start-up company time to build up its brand, as it may have no brand initially given that it is essentially an unknown entity. Copyrights and patents can give the company protections from stronger brands, and these protections give the start-up time to build up its reputation and brand.

Trademarks are therefore important to companies and companies may sue other companies for infringement of their trademark in a particular country. Countries vary in their protection of trademarks, and while most countries are rigorous in dealing with violations of patent law or copyright law, some countries are less stringent with violations of trademarks and may make very small awards. A company that is successful in a trademark lawsuit may be awarded damages for trademark violation, and the penalties could include the actual damages, an award of all or part of the profit earned by the violator of the trademark, and punitive damages as punishment for the transgression.

There are different types of trademarks such as the common trademark to protect products, the service mark to protect services, and the certification mark to protect certifications. There are global treaties (e.g., the Paris Convention on global brand recognition) that apply from country to country related to trademarks, but there is no real international trademark law.

10.4.1 Legal Aspects of Trademarks

Trademark law is local law and country specific with significant variations between countries (or even within a country), and registration is required in every country that is important to the business. It is important to register the trademark and this often requires detailed paperwork to be completed, and the payment of an annual fee to keep the trademark registered in the country. In some countries (e.g., the US) it is not required to register your trademark in order to defend it in a lawsuit, but in some other countries you have zero rights if you fail to register. Registration is required in most countries if you wish to sue another person for infringing your trademark (e.g., if you do not register your trademark in China then someone else may do so resulting in zero legal rights for the plaintiff). Another words, registration is important and it is not that expensive to do.

The registration of a trademark is not mandatory in some countries as the rights to a trademark may be granted based on its use. Common law trademarks exist in some countries (e.g., UK), and these can exist without registration. These are the only form of trademark that existed in the past, but they are not valid in some countries (e.g., it is not possible to sue for trademark violation in the US unless it is registered). A registered trademark is indicated by $^{®}$, whereas the symbol ™ in common law indicates a trademark for goods that may be unregistered, with the symbol ^{SM} indicating a trademark for services that may be unregistered.

There is a distinction between the trademark rights from the *first to register* the trademark versus the *first to use*. The fact that you are the first party to register the trademark may not be sufficient for it to be granted, as another party may already be using the trademark. In such a case the party that is already using it has priority due to usage and has the right to continue using it. Some countries such as the United States award the trademark rights to the first party to use it, whereas others such as China award priority and rights to the party that first registers the trademark. The implication for a company in a first to file country is that if the company does not file first then it will have a major challenge in protecting its brand in that country.

It is important that the trademark is being used, or that there is good faith and intent to use it over time (the product may be in development). Otherwise, if the trademark is not being used and a second party applies for the same trademark, then the second party is in a position to argue that the first party is showing a lack of good faith, and is not intent on using the trademark. The right to exclusive use of a trademark derives from its appropriate and subsequent use in the market place, and so while a first party might get a trademark it cannot keep it unless it is being used, and continues to be used.

A trademark may potentially last forever, as it does not expire after a fixed period of time provided that it continues to be used. For example, brands such as "Coca Cola" or "Budweiser" have been in existence for over 100 years. However, trademarks may be lost over time through non-usage, or if they become generic or deceptive.

Distinctive words are easier to register, whereas generic words are more difficult to register. It is difficult to get approval for a trademark where there is likelihood that it would cause confusion with an existing trademark. There are potential trademark issues in cyberspace, and so it is also important that a company registers its domain names in all the countries in which its e-commerce site will be operating. Otherwise, there is a danger that the same name may already be used within the country, which might prevent e-commerce from taking place, as access to the website is blocked.

A judge is the deciding party who determines whether there is a trademark violation when one party takes a lawsuit against a second party for trademark infringement. The judge will determine how similar the trademarks are, as well as how similar the goods or services are, and the likelihood of confusion between the trademarks, as well as evidence of confusion between the products. The judge will examine when the trademarks were first registered and first used, and will take into account the strength of the trademarks. The judge will determine whether it was the deliberate intent of the second party to choose the name due to its similarity to the first party's brand, and whether there is direct or secondary evidence of confusion between the products or the likelihood of confusion.

The second party will make a defence against the infringement, and the legal defence may be that its use of the trademark is fair and reasonable. It may argue that it first registered or used the trademark, and there is no confusion between them, and that consumers can adequately distinguish between the products, or that the products are dissimilar.

The award for a successful lawsuit may be monetary damages to the first party to account for lost profits and damages (possibly including punitive damages as punishment) and court costs, as well the possibility of an injunction being made against the second party to stop it from selling its products in the particular jurisdiction. The second party may appeal the judgment or size of the judgment awarded to the first party.

Trademark trolls are individuals or companies that register a trademark or company name or local domain name of an e-commerce site in another country (in a first to file country) before the original owner of the trademark, or the e-commerce site. The goal of the trademark troll is to profit from this either by demanding a payment to transfer the rights to the original company, or by threatening to take or actually taking legal action for infringement of their registered trademark. Start-up companies need a strategy to deal with trademark trolls to ensure that they are able to expand internationally without being adversely affected by trolls.

Companies invest in their brands and take lawsuits as part of their strategy to defend and protect their brand, and so trademarks may be used as a strategic weapon to defend their market share against new entrants. The goal is to defend against new brands that may damage their existing brands, and so the net effect is to slow down and weaken smaller rivals and preserve market share.

10.4.2 Case Study—Trademark Dispute Apple Corps Versus Apple Computer

Apple Corps Ltd. (informally known as Apple) was founded in London in 1968 by the members of the Beatles, and with its chief division being Apple Records, which was the record label for the Beatles and was founded in the same year.

Apple Computer Company (today called Apple Inc) is a well-known American corporation that was founded by Steve Wozniak and Steve Jobs at Cupertino, California in 1976. It has made major contributions to the computing field, and it is renowned for its design of innovative products and services. It has designed and developed world-class computer hardware and software such as the Apple and Macintosh computers; dazzling smartphones such as the family of *i*Phones; and the *i*Pad tablet computer. It is the largest music vendor in the world, as its online *i*Tunes digital media store has over 35 million songs and 45,000 films. Its App store allows users to download applications developed for Apple's iOS operating system to their *i*Phone or *i*Pad. Apple Inc. is discussed in more detail in Chap. 5 of [4].

Apple Corps filed a lawsuit against Apple Computer in 1978 for trademark infringement and the lawsuit was settled in 1981. The outcome was a financial settlement paid to Apple Corps, and an agreement that Apple Computer would not enter the music business, and that Apple Corps would not enter the computer business.

Apple Computer added MIDI and audio recording capabilities and an advanced sound chip to its computers in the late 1980s, and this led Apple Corps to sue again in 1989 alleging violation of the 1981 agreement. This resulted in Apple Computer terminating its work in the multimedia field at that time.

Apple Corps took another action against Apple Computer in 1991 following the addition of a sound system to the Macintosh operating system by an Apple Computer employee. A further financial payment was made to Apple Corps and under the terms of the agreement Apple Corps had the rights to use "Apple" on any creative works that included music, whereas Apple Computer had the right to use "Apple" on goods and services used to run, play or deliver such content but not on content delivered on physical media.

Apple Corps sued Apple Computer in 2003 for breach of contract for using the Apple logo in the creation of the iTunes music store. Apple Corps had rejected an offer from Apple Computer for a financial payment for use of the "Apple" name prior to the launch of the iTunes store. However, the court ruled in favour of Apple Computer in 2006, and ruled that no breach of the trademark agreement had been demonstrated, as Apple Computer's use was covered under the existing agreement.

Apple Corps indicated that it planned to appeal the judgment, but in 2007 an agreement was reached between the parties, which allowed Apple Inc. to own all of the trademarks related to "Apple". The agreement allowed the licensing of some of the trademarks back to Apple Corps for their continued use. This agreement and the associated financial settlement ended the trademark dispute between both companies.

10.5 Review Questions

1. What is a patent?
2. What is the process for obtaining a patent?
3. Explain how the ABC was ruled to be the first electronic digital computer.
4. What is a copyright and what rights does it give?
5. What is copyright infringement?
6. Explain the main parts of the Digital Millennium Copyright Act.
7. What is a trademark?
8. Explain the distinction between first to register and first to use in trademark law.
9. What are trademark trolls and how do they operate?
10. What is a trade secret? How does it differ from a patent?

10.6 Summary

Intellectual property law deals with the rules that apply in protecting inventions, designs and artistic work, and in enforcing such rights. The inventor is generally granted exclusive rights to the invention for a defined period of time, which provides an incentive to develop creative works that may benefit society.

The main forms of intellectual property are patents, copyright and trademarks. Patents protect innovative ideas and concepts, and give inventors exclusive rights to their invention for a specified period or time. The invention must be novel and more than an obvious next step from existing technology, and the patent needs to be filed at the Patent Office.

A copyright applies to original writing, music, and other original intellectual and artistic expressions. It protects the expression of the idea and not the underlying idea itself. Copyrights provide exclusive rights to making copies of the expression, where the ways of expressing ideas is copyrightable. The term "fair use" refers to the permitted limited use of copyrightable material without acquiring permission from the copyright owner.

A trademark protects names or symbols that are used to identify goods or services, and their purpose is to avoid confusion and to help customers to distinguish one brand from another.

References

1. G. O' Regan, *Giants of Computing* (Springer, Heidelberg, 2013)
2. G. O' Regan, *World of Computing* (Springer, Heidelberg, 2018)
3. R. Foroohar, *Don't be Evil. The Case Against Big Tech* (Allen Lane, 2019)
4. G. O' Regan, *Pillars of Computing* (Springer, Heidelberg, 2015)
5. R. Stallman, *Free Software, Free Society* (Free Software Foundation, Inc., Boston, 2002)

Legal and Ethical Aspects of Electronic Commerce

11

Abstract

This chapter discusses e-commerce and e-commerce law, where e-commerce is a way to conduct business online. It involves listing products for sale in a catalogue format on a website, and there are also legal issues to consider. E-commerce is different from traditional business in that the buyer and seller do not physically come together in the market place to perform the transaction, and they may even be in different countries with distinct laws. There may be a greater risk of fraud or loss with e-commerce than with traditional commerce.

Keywords

Internet • World Wide Web • E-signatures • Payments • Website • B2B • B2C

11.1 Introduction

Tim Berners-Lee invented the World Wide Web in the late 1980s, and he realized that the Web offered the potential to conduct business in cyberspace, rather than the traditional way where buyers and sellers come together to do business at the market place. The success of the Web led to modern electronic commerce, and the required technologies (e.g., electronic funds transfer, secure socket layer, electronic data interchange, online transaction processing, and so on) were later developed. E-commerce is a way to conduct business online, and it includes many different services such as shopping, booking flights and hotels, banking and financial services, entertainment, making online payments, and so on.

The growth of the Web was phenomenal, and exponential growth rate curves became a feature of newly formed Internet companies and their business models. It was predicted that the new web-based economy would replace traditional bricks and mortar companies, and it was expected that most business would be conducted

Fig. 11.1 E-Commerce transaction

over the web, with traditional enterprises losing market share and going out of business (Fig. 11.1).

Exponential growth of e-commerce companies was predicted, and the size of the new web economy was estimated to be in trillions of U.S. dollars. The United Nations Conference on Trade and Development (UNCTAD) estimated that the size of the global e-commerce economy was over $25 trillion in 2018, and that this was equivalent to 30% of global GDP [1]. The value of B2B e-commerce was estimated to be $21 trillion, with B2C e-commerce estimated to be $4.4 trillion. Forecasts for B2B and B2C for various regions and segments are in [2].

New companies were formed to exploit the commercial opportunities of the Web, and existing companies developed e-business and e-commerce strategies to adapt to the brave new world. Companies providing full e-commerce solutions began to sell products or services over the web to either businesses or consumers. These business models are referred to as Business-to-Business (B2B), Business-to-consumer (B2C) and Business to Government (B2G). E-commerce web sites have the following characteristics (Table 11.1):

The success of the World Wide Web was phenomenal and it led to the formation of many "*new economy*" businesses. These businesses were conducted over the web, and included the on-line bookstore, Amazon; and the on-line auction site, eBay. Amazon initially sold books only, but it is now a major e-commerce site that sells a vast collection of consumer and electronic goods, whereas eBay brings buyers and sellers together in an on-line auction space.

Some of these new technology companies were successful and remain in business. Others were financial disasters due to poor business models, poor

11.1 Introduction

Table 11.1 Characteristics of E-commerce websites

Feature	Description
Well-designed and easy to use	Usability of the web site is essential, as otherwise the web site will not be attractive to users
Catalogue of products	The catalogue of products details the products available for sale and their prices
Shopping carts	This is analogous to shopping carts in a supermarket, where items selected for purchase are placed in an electronic shopping cart
Payments	Once the user has completed the selection of purchases there is a checkout facility to arrange for the purchase/payment of the goods
Security	Security of credit card and financial information is a key concern for users of the web site, and users need to have confidence that their financial and personal data will remain secure
Order Fulfilment/Order Enquiry	Once payment has been received the products must be delivered to the customer. There are several order fulfilment models

management, and poor implementation of the new technology. Some of these technology companies offered an Internet version of a traditional bricks and mortar company, with others providing a unique business offering. For example, eBay was offering a new service that was distinct from traditional auctioneering.

Jeff Bezos founded Amazon in 1995 as an on-line bookstore. Its product portfolio has expanded in a major way to just about everything, and it also offers cloud-computing services. Its initial focus was to build up the "Amazon" brand throughout the world, and it initially sold books at a loss by giving discounts to buyers to build market share. It was very effective in building its brand through advertisements, marketing and discounts.

The initial public offering of Netscape in 1995 demonstrated the incredible value of these new Internet companies. The company had planned to issue the share price at $14, but it decided at the last minute to issue it at $28. The share price reached $75 later that day. This was followed by what became the dot com bubble, where share prices of e-commerce companies reached astronomical levels. Reality returned to the stock market when it crashed in April 2000, and share values returned to more realistic levels.

The vast majority of these early Internet companies were losing substantial sums of money, and few expected to deliver profits in the short term. Financial instruments such as the balance sheet, the profit and loss account, and the price to earnings ratio are generally employed to estimate the value of a company. However, investment bankers at the time argued that there was a new paradigm in stock market valuation for Internet companies. This paradigm suggested that the potential future earnings of technology companies be considered in determining their value, and this was used to justify the high prices of shares, as frenzied investors rushed to buy these over-priced and over-hyped stocks. Common sense seemed to play no role in rational decision-making, and the herd mentality associated with bubbles in asset prices seemed to be the main vehicle employed in decision-making.

The fall of the indices was equally as dramatic especially in the case of the NASDAQ. It peaked at 5000 in March 2000, and fell to 1200 (a 76% drop) by September 2002. It had become clear that Internet companies were rapidly going through the cash raised at their IPOs, and analysts noted that a significant number would be out of cash by the end of 2000. Therefore, these companies would either go out of business, or would need to go back to the market for further funding. This led to questioning of the hitherto relatively unquestioned business models of these Internet firms.

Funding is easy to obtain when stock prices are rising at a rapid rate. However, when prices are static or falling, with negligible or negative business return to the investor, then funding dries up. The actions of the Federal Reserve in rising interest rates to prevent inflationary pressures also helped to correct the "irrational exuberance" of investors.

11.1.1 Advantages of E-Commerce

E-commerce has led to major benefits for consumers in that it often leads to lower prices, as less staff are required to manage an online store when compared to a standard bricks and mortar store. Often, there are savings through the use of automated inventory management as well as warehousing, and this allows e-commerce business owners to pass on these savings to their customers.

There are price comparison websites that allow consumers to shop around to compare prices and obtain the best possible deal. Online shopping is attractive to consumers, as e-commerce sites are available at all times (i.e., 24 × 7 × 365), which allows customers to shop at their convenience without going to a physical store. The growth in e-commerce has led to a massive increase in the choice of products available online, and the consumer is not restricted to the specific products that are available in their locality.

The advantages of e-commerce to a business is that it often leads to higher profit margins for the online company, as there is a reduction in operating costs such as rent, staff, heating and electricity, warehousing and inventory management. Further, a physical store is subject to physical limitations such as the number of customers that may be in the store at any one time, whereas there is potentially no limit on the number of customers that may be shopping on an online store. This means that there is major potential for growth of on-line stores into new geographical areas, subject to satisfying the legal and taxation requirements in the various jurisdictions.

Finally, e-commerce sites gather a lot of data to understand consumer behaviour on the online site, and this data may be used to understand customer-buying habits and to improve the user experience. The successful implementation of improvements based on the data gathered may lead to an increase in sales.

11.1.2 Disadvantages of E-Commerce

There are several disadvantages of e-commerce when compared to traditional brick and mortar companies such as no one can purchase from an online store if the website crashes and is unavailable for a period of time. It is therefore essential that the website is hosted on the right platform, and that the platform is reliable and scalable to support future growth in traffic. It is essential that the website is secure and protects the personal and financial information of its customers, as any security breaches will lead to a loss of confidence in the site.

In traditional retail, customers are able to try on products to ensure that they fit and are suitable. This is not possible with traditional websites but augmented reality may allow a more interactive experience for customers in the future. Further, there is a social dimension to traditional retail, which is lacking in the online world.

In traditional retail the customer may purchase the product and bring it home immediately, whereas in the online world there are delays in receiving the product as the shipping may take weeks. Further, if the online purchase is made in another country it may be subject to customs duties and additional charges on arrival in the country.

11.2 E-Commerce Law

Electronic commerce refers to the marketing and sale of products or services over the World Wide Web, and it often involves listing products for sale in a catalogue format on a website. In some cases, such as the sale of software, music or an electronic book, the actual sale and delivery can take place online, but in general there is a requirement for the physical delivery of the goods. An e-commerce website needs an attractive design for its users as well as being easy to use, and the electronic site includes the catalogue of products, a shopping cart and a payment system.

Further, the implementation of the electronic commerce site must satisfy all applicable legal and regulatory requirements, and there may be both general and specific regulations. The general regulations with respect to electronic commerce include:

- Data protection legislation such as GDPR
- European Digital Service Act (DSA) for EU countries
- Distance selling legislation (e.g., cooling off period)
- Terms and conditions of contract
- Company address and registration

There may be additional regulatory requirements for the particular sector of the business, as for example, the banking sector, which would have additional regulatory requirements relating to the security of financial transactions.

The rise of electronic commerce has raised legal questions on the extent to which online agreements are enforceable, which country's laws are applicable, and consumer protection rights. There is an increased risk of fraud with electronic commerce, and as activities are not conducted face-to-face there are practical difficulties in pursuing claims in foreign countries. And so, it is essential that there are laws that support electronic commerce as well as providing protection to consumers.

E-commerce is different from traditional business in that the buyer and seller do not physically come together in the market place to perform the transaction, and there is a greater risk of fraud or loss with e-commerce than with traditional commerce. Electronic information is legally valid and a contract may not be denied validity due to the fact that it is in electronic form. That is, contracts made over the Internet are as legally binding as contracts made face-to-face or in written form, and so contracts can be made as freely over the Internet, email and other electronic means as via face-to-face or written communication.

The usual method of authenticating consent to a written (traditional) contract is with a physical signature to confirm consent and agreement. However, a physical signature is not possible for electronic commerce transactions, and instead electronic signatures are used for authentication. An electronic signature is data in electronic form that is associated with other electronic data as a means of authenticating the originator. Another words, this may take the form of associating data with other data, such as a name at the end of an email that originates from that person as a way to authenticate the originator. Another option would be to type the name of the signer in the signature space, and paste in the scanned version of the signature. Another form of authentication that is used in the banking sector is password authentication, and there are also several strong authentication methods available that are used to authenticate a user's identity and to prevent attacks or scams.

Websites collect; store, and process personal information from consumers, and so online websites must protect personal information and satisfy the data protection laws in the country in which they are operating. Data privacy laws such as GDPR regulate the holding of personal information in electronic form, where this legislation is intended to protect the privacy of individuals and to prevent personal data from being misused. This means that personal information must be obtained lawfully, and individuals must be informed about the purpose of the data collection and give their consent, and the individual must approve any changes to the way in which the data is used.

The European Digital Service Act (DSA) became part of European law in 2024, and it is applicable to all member states in the European Union. The goal of this legislation is to protect digital space against illegal content as well as protecting the fundamental rights of users, and it is applicable to social media companies and the online marketplace. Platforms will be held accountable (in Europe) for their role in disseminating harmful content, and it lays down obligations on platforms and control measures.

The terms and conditions by which goods are offered for sale need to be compliant with consumer protection laws. The terms and conditions cover the terms of sale as well as how the goods and services are advertised and marketed, as well as the rights of consumers both before and after a sale.

11.2.1 Website and Hosting

The online trader may not own all the rights to its website content (e.g., the copyright to images, music or software used, or trade mark protected material), and so the trader must obtain permission (this may require licensing) to exploit such materials. Otherwise, the trader could be subject to legal action from the content owner for infringement of their rights.

An online trader must not post any content that could be deemed defamatory, and an online trader must not post content which interferes with any individual's personal data or privacy rights under data protection law, and other applicable law (e.g., incitement to hatred).

The online trader should publish the name, address and contact details of the website operator, and this may include the place of registration of the company and its registration number and the address of the company's registered office.

The online trader is responsible for the accuracy of the content of the website, and in certain situations (e.g., unlawful activity such as copyright infringement) the Internet Service Provider (ISP) may act to remove or disable access to the website and its contents. Otherwise, the ISP could be considered to be a party to unlawful activity, and so, the ISP may, without permission, shut down a website, remove content, or disable linking as it may be required by law to take such action.

The online trader must post accurate product information on its website, and needs to be aware of the prohibition of unfair and misleading commercial practices. Under consumer protection law an advertisement is misleading if it has false or deceptive information, and this could include false claims about a product or leaving out important information. Further, the consumer should exercise caution with product reviews, as some are not genuine (fake reviews), and businesses must ensure that any reviews that they show are genuine.

11.2.2 Registering a Domain Name

It is essential that the domain name for the e-commerce website is registered in the countries where the website will be used. There is always the danger of *cyber squatting*, where others have already registered the Internet domain name. These other individuals or companies hope to profit from selling the domain name to the owner of the trademark at an inflated price.

That is, the goal of the cyber squatter is to profit from registering the domain name by demanding a payment to transfer the rights to the original company. There are laws against cyber squatting in some countries, where victims can take legal action to deal with the situation, but a company may be exposed to such practices in other countries. Hence, it is important to register the domain name of the website as early as possible in all countries where it will be used.

This means that it is important to choose a domain name carefully, and the domain name should be available and should not infringe upon another person or the trademark rights of another entity. Further, the name should not be confusing or easily associated with another business. We mentioned in Chap. 10 that the rights to a trademark in China are awarded on a first to file basis rather, than a first to use basis, and so it is essential to file the domain name as early as possible in China to prevent others from cyber squatting.

11.2.3 Privacy Policy on Website

Websites collect and store information from the people who visit it, and under the European GDPR regulations users must be informed of the information that is held about them, as well as how the information will be used. The e-commerce site will need a privacy policy to inform the users of their privacy rights, and how the law will protect them. Users have several protections under GDPR, and their personal data must be processed lawfully, fairly and transparently. These rights include:

- User consents to the processing
- Processing is necessary to carry out the contract with user.
- Processing is necessary due to legal obligations.
- Processing is necessary for public interest
- Personal data is accurate and kept up to date
- User consent is required for sensitive data.

Under GDPR users have rights to enquire into whether the e-commerce website is processing their personal data, and if so, they have a right to be given a description of the personal data. Further, they have the right to ask about the purpose for which the personal data is being processed, as well as who the personal data has been disclosed to. There should be a link on the website to the data protection policy, and the user must confirm that they have read the policy.

Users also have rights with respect to cookies that are downloaded to their device when they visit a website. Cookies are used to improve the user experience as well as building a profile of the user, and these are pieces of data sent by websites that remain stored on a user's computer to help with future web browsing. The e-commerce website must have a cookies policy informing the user that cookies are being used when they access the website, and so users must be informed and given information or a link to information that explains their use. The information explains the purpose of cookies before they are set on the user's browser.

Users have the right to complain to the Data Protection Commissioner (in GDPR jurisdictions) if the electronic commerce website fails to modify incorrect or incomplete information about them.

11.2.4 Terms and Conditions

The terms and conditions of use of a website governs how it is used, and the applicable terms and conditions for selling goods and services on the website. The terms of conditions must satisfy the relevant legal requirements for conducting commerce over the Internet, as well as satisfying privacy laws such as the European GDPR regulations and the European Digital Service Act for protecting the fundamental rights of consumers. The customer must confirm that they have read the terms and conditions as well as the privacy policy.

The terms and conditions of a website that offers goods or services for sale need to be compliant with consumer protection laws, which cover the terms of sale, as well as how the goods or service are marketed or advertised, and the rights of consumers before and after a sale.

The applicable terms and conditions of sale and supply are generally incorporated into customer contracts, and a link is provided to the terms so that the customer may locate them easily. It is normal to have a step in every sale where the customer confirms that they have read the terms and conditions of supply and the privacy policy. The user clicks on the "I Agree" or "Click to Agree" to confirm their acceptance of the terms and conditions of the contract.

11.2.5 Contracts

A contract between two parties consists of an offer and an acceptance, and contracts may be formed electronically between both parties, and they may be concluded electronically. That is, an electronic contract is an agreement that is drafted, signed, and executed completely online, and it consists of an electronic offer by one party, and an electronic acceptance of the offer by the other party.

An electronic contract is similar and drafted in the same manner as a traditional paper-based contract. The offer and acceptance is communicated electronically unless otherwise agreed by the parties, and such contracts are subject to the core principles of contract law of the particular jurisdiction. The fundamental principle is that for a contract to arise, there must be an offer and acceptance between both parties.

The buyer browses the available goods and services on the seller's website, and chooses what he/she wishes to purchase. The customer makes an offer by placing an order for the selected products in the shopping cart for payment, and the seller acknowledges the offer (usually via an email), and the seller's acceptance of the offer (and the confirmation means that there is now a contract between both parties) is generally given by email (possibly an automated email).

The remedies available for a breach of an electronic contract are largely the same as for conventional offline contracts. These could include the right to terminate the contract, the right to cancel (rescind) the contract where there is a material breach of the contract, the right of restitution (i.e., restoring to a former condition), or damages. There may be additional remedies available to a consumer where certain terms implied by legislation in the jurisdiction are breached, and where there are regulations and protections of consumers including the repair/replacement of a faulty product or a reduction in price.

Arbitration is another way in which a dispute may be settled between parties that avoids the need for litigation. However, should the parties be unable to resolve their dispute, then there are legal remedies available to deal with the breach of contract. The party that has suffered loss as a result of the breach of contract may sue the other party for financial compensation. However, it is desirable to avoid litigation due to the complications of different jurisdictions and legal systems, and usually both parties will aim to resolve the dispute amicably.

11.2.6 Sale of Goods or Services

Consumers must be given specified information before they order, and once a customer places an order they must receive a confirmation of the order. Under European consumer rights law, consumers have a 14-calendar day cooling off period should they wish to withdraw from the contract without giving a reason or if they are not satisfied with the product. The e-commerce website should inform the user of their rights, and, should the consumer decide not to proceed during the cooling off period, he/she may need to bear the cost of returning the product to the online business.

The invoice supplied to the customer must include the contact details of the trader, as well as the address for complaints. The goods should be described as well as the associated price and shipping costs. All relevant contract information such as delivery periods, payment methods, cancellation options need to be specified in the contract, and the customer must receive notification of the purchase either electronically or other means within 24 h.

There are responsibilities on the seller to ensure that the supplied goods are as described, and that the goods are fit for purpose as well as being safe and free from defects and are of a satisfactory quality.

There are remedies available to the customer when the goods do not conform to the terms of the contract, and the remedies (depending on the circumstances) could include a partial or full refund of the payment made, repairs or replacements of the goods, or cancellation of the contract.

11.2.7 Online Marketing and Advertisement

All marketing communication on the website or elsewhere should be legal and truthful, and should not include anything that is likely to cause offence (e.g., in terms of race or religion). Advertisements or marketing material should not be misleading, and the claims made in an advertisement should be accurate. Under consumer law an advertisement is misleading if it has false or deceptive information, or if important information is omitted.

The electronic commerce website may be liable should any misleading descriptions arise that cause confusion or misrepresent the characteristics of the goods or services that are offered for sale in an advertisement. An advertisement may be subject to different interpretations by consumers in different jurisdictions leading to potential liability under consumer protection law. An online advertisement should be truthful and not misleading.

11.2.8 Jurisdiction

There are several legal issues related to e-commerce in relation to jurisdiction of an electronic transaction, including which country's courts should have jurisdiction in resolving disputes that arise? Should it be in the buyer's country? Should it be in the country where the electronic commerce firm is located? Further, if a court decides that it has jurisdiction in the matter, should it apply the law of its own country or that of another country? What is the basis on which courts and regulators claim jurisdiction?

Other relevant questions include how should consumers be protected in e-commerce transactions, and who is responsible for regulating Internet activity? How can an individual who resides outside the relevant jurisdiction obtain a just resolution to a dispute that arises with an e-commerce company? The European Digital Service Act aims to protect the fundamental online digital rights of consumers in the European Union.

The jurisdiction of an electronic commerce transaction may affect consumer rights and consumer protection.

- Which jurisdiction should apply?
- How is jurisdiction decided?
- Which laws should be followed?
- Where should action be taken?

The jurisdiction of a transaction is relevant to consumer protection and the enforcement of rights in dispute resolution, and it is desirable to find a resolution that does not involve the complexities of international litigation.

11.2.9 E-Signatures

An electronic signature refers to data in electronic format that is attached to or logically associated with other electronic data, and which serves as a method of authenticating the originator. It may include a scanned electronic signature, where the scanned electronic signature is uniquely linked to the signatory, and is capable of uniquely identifying the signatory. The electronic signature is linked to the data to which it relates in such a manner that any subsequent change of the data is detectable.

Many contracts may be concluded using electronic signatures, but the execution of certain documents (e.g., wills, trusts, affidavits and sworn declarations) often requires traditional signatures. For example, a contract for the sale of land may be concluded electronically, but often the deed of conveyance may require a traditional signature.

11.2.10 Payment Systems

There are strict rules governing electronic payments that include rigorous security requirements for electronic payments, as well as protection of the consumers' financial data. Credit and debit cards are the most common payment methods for electronic commerce payments, and the website must be secure to ensure that personal financial information is not compromised.

11.2.11 Taxation

There may be taxes due from the sale of goods and any taxes are payable in the relevant jurisdiction. There may also be duties and taxes to be paid on the shipping of the goods to the final destination.

11.2.12 Consumer Protection

Consumers must be given specified information before they order, and once a consumer places an order they must receive a confirmation of the order, and they generally have a cooling off period (in the EU consumers have 14 days) where they may cancel the contract without giving a reason.

The terms and conditions of a website that offers goods or services for sale need to be compliant with consumer protection laws, which cover the terms of sale, as well as how the goods or service are marketed or advertised, and the rights of consumers both before and after a sale.

Under consumer protection law an advertisement is misleading if it has false or deceptive information, and this could include false claims about a product or leaving out important information.

11.3 Ethical Electronic Commerce

Ethical electronic commerce is concerned with ways of providing a positive online experience to consumers, as well as improving business efficiencies to benefit both consumers and the wider environment. Many consumers wish to purchase more responsibly in a way that causes less harm to the environment and other people, and so ethical electronic commerce includes sustainable, person-first, and green business practices.

Sustainable electronic commerce involves adopting practices that minimise the impact on the environment, and often this involves redesigning business processes as well as improving transportation, logistics and packaging. Consumers have become interested in sustainable and environmentally friendly packaging, with the less packaging the better, as most packaging ends up in a landfill.

A people-first policy is where the business puts the interest of the consumer first even at the expense of profits. This includes all people involved in the process of manufacturing goods, including both offshore and onshore workers and ensuring that they have safe working conditions and are paid a fair wage.

Greenwashing is where a business makes a false claim regarding their business practices being conducted ethically, thereby giving the impression of an ethical brand. That is, the company provides misleading information that gives the impression to people that it is doing more for the environment than it really is.

11.4 Review Questions

1. What is e-commerce?
2. Explain the difference between B2B, B2C and B2G.
3. Describe the characteristics of an e-commerce website.
4. What are the advantages and disadvantages of e-commerce?
5. What rights do the cooling off period give the consumer?
6. What rights does the consumer have if the product is faulty?
7. Explain the terms and conditions of use of a website.
8. Explain the responsibilities of the seller when selling products/services on line.

11.5 Summary

E-commerce is a way to conduct business online, and it includes many different services such as booking flights, carrying out financial services such as banking and insurance, purchasing goods and services, and so on. It often involves listing products for sale in a catalogue format on a website, and the user selects items for purchase and makes an electronic payment.

E-commerce differs from traditional business in that the buyer and seller do not physically come together in the market place to perform the transaction, and they may even be in different countries with distinct laws. There may be a greater risk of fraud or loss with e-commerce than with traditional commerce.

Electronic information is legally valid, and an electronic contract consists of an electronic offer and electronic acceptance to be valid. The terms and conditions by which goods are offered for sale need to be compliant with consumer protection laws, and these cover the terms of sale as well as how the goods and services are advertised and marketed, as well as the rights of consumers both before and after a sale. There are responsibilities on the seller to ensure that the supplied goods are as described, and that the goods are fit for purpose as well as being safe and free from defects and of a satisfactory quality.

There are remedies available to the customer when the goods do not conform to the terms of the contract, and this could include a partial or full refund of the payment made, repairs or replacements of the goods, or cancellation of the contract.

References

1. *Global e-commerce hits $25.6 trillion.* UNCTAD Press Release, PR/2020/007. April 2020. https://unctad.org/press-material/global-e-commerce-hits-256-trillion-latest-unctad-estimates
2. E-commerce Market Size, Share and Trends Analysis Report by Model Type (B2B, B2C), by Region and Segment Forecasts, 2020–2027. May 2020. GVR-4-68038-684-4

Computer Crime

12

Abstract

Computer crime (or cybercrime) is a crime that involves a computer and a network. The computer may be the vehicle by which the crime is conducted.

Keywords

Computer crime • Scam • Malware • Hacking • Cyberextortion • Ransomware • Cybersecurity • Phishing • Trojan horse

12.1 Introduction

Computer crime (or cybercrime) is a crime that involves a computer and a network. The computer may be the vehicle by which the crime is conducted, or it may be the target of the crime. It is common to encounter dangers in some streets or neighbourhoods of major urban areas, and such dangers need to be managed appropriately. Similarly, the Internet has dangers with hackers, scammers, and web predators lurking in the shadows, and so the user needs to be cautious in her actions and needs to manage the situation.

A hacker may be accessing a computer resource without authorisation with the intention of committing an unlawful act. The hacker's activities may be limited to *eavesdropping* (listening to a conversation), or it may be an active *man-in-the-middle* attack. The latter is where the hacker may possibly alter the conversation between two parties with the intention of committing an act of theft, or some other malicious action on an unaware victim.

There is a popular story (it seems plausible) that suggests that one of the earliest computer crimes that occurred in the United States was in the late 1950s. It was a financial crime that involved a programmer opening up a bank account, and then writing a program that funnelled all of the fractions of a cent from the thousands of accounts in the bank into his account (i.e., the program involved the alteration of bank records to embezzle fractions of a cent from the accounts in the bank). And while the amount taken from each account was miniscule, the sheer volume of accounts meant that a large amount was stolen from the bank.

One of the earliest Internet attacks was in 1988, when a graduate student from Cornell University (Robert Morris) released a program on the Internet (an Internet Worm later termed the "Morris Worm"), which exploited security vulnerability in the mail software to automatically replicate itself locally and on remote machines. It affected lots of machines and effectively shut down the Internet for 1–2 days. Morris developed the malicious software at Cornell and released it by hacking into a MIT computer. Morris had intended it as a harmless experiment, but the virus rapidly spread out of control. Morris was prosecuted in 1989 under the 1986 Computer Fraud and Abuse Act that prohibited unauthorised access to protected computers. He was fined, given probation and ordered to complete 400 h of community service [1].

The legal system evolves as technology evolves and an accepted legal principle is that a person cannot or should not face criminal punishment for an act that was not a criminal offence at the time when he or she committed the act (*nullum crimen sine lege*). Another words, the principle says that there is no punishment without law, and this protects perpetrators from punishment outside of the law. That is, there was very little law dealing with computer crime in the early days of computing, and so there was minimal punishment of offenders, as the probability of conviction in the absence of law was very low. Prosecutors attempted to use the existing law in an attempt to bring perpetrators to justice, but often there were loopholes that could be exploited by the defence.

The number of computers and computer users was extremely small in the early days of computing (mainly large mainframes), and the introduction of time-sharing systems in the 1960s was a way to share scarce computer resources among several users. This led to an increase in the number of computer users, but the vast majority of the population had no access to computer technology. There were no personal computers, laptops, tablets and smartphones. Most computer crimes that occurred involved some form of theft committed internally, with unauthorised access regarded more as an abuse of computer resources rather than a crime. This was since computer resources were still quite rare at that time, and so unauthorised access to computer resources was more socially acceptable than it is today. In the modern age, computer resources are everywhere with individuals often having one or more computers or laptops, and so there is no justification (it is not acceptable) for unauthorised individuals to gain access to computer resources.

Most developed countries introduced laws criminalizing computer crime from the early 1970s, and these laws have evolved as technology evolves. Often new problems arise from the introduction of new technology, and so the legislature and legal system needs in a sense to catch up with new technology, to ensure that technology serves society. For example, the introduction of the Internet and World Wide Web has transformed the world, but it also had the unintended consequences of transforming computer crime onto the global stage, as billions of users are now accessing web sites all around the world. This has created a whole new set of problems and challenges for the legal system to deal with. The European Union has been active in developing appropriate legislation for the digital age, and its contributions include legal frameworks such as GDPR and DSA.

12.2 Types of Computer Crime

Today, more and more individuals and companies are on line, and networking systems and computers have become quite complex. There has been a major growth in attacks on businesses and individuals, and so it is essential to consider computer and network security. The Internet was developed based on trust with security features added later as a response to different types of attacks.

There are several threats associated with network connectivity such as *unauthorised access* (a break-in by an unauthorised person), *disclosure of sensitive information* to people who should not have access to the information, and *denial of service* (DoS), where there is a degradation of service that makes it impossible to access the web site and perform productive work. Table 12.1 presents some examples of computer crimes:

There are many possible computer crimes varying from attacks that lead to defacement of web sites, to bank fraud and the theft of credit card numbers, the sending of hoax (scam) letters that are designed to deceive and steal from the user; phishing emails that appear to come from legitimate parties, but links to a site that is different from the one that the user expects to go and attempts to deceive and steal from the user. Other crimes include the misuse of packet sniffing tools to intercept packets on the network (especially on a public WiFi network that lacks security protocols) thereby gaining access to sensitive data such as user-ids and passwords. We discuss some of these cybercrimes in more detail below.

12.2.1 Scams

A *scam* is a scheme where one party is intent on deceiving another for financial gain, and it generally involves persuading the victim to part with his/her money. The scam may be extremely convincing, and the communication may appear to come from a legitimate source. A scam is often sent by a hoax email that is designed to deceive, and fraud the email recipient.

For example, the victim may receive an email from a party that appears to be their bank, and advise them that they are overdrawn on their current account, or that there is a suspicious transaction on their account. They are requested to login with the link provided in the email, and once they have clicked on the link and provided their financial details to what appears to be their bank, they suffer immediate financial loss.

Phishing is an attempt to obtain sensitive information such as usernames, passwords and credit card details with the intention of committing fraud. The scam may take the form of a phishing email designed to persuade the victim to reveal

Table 12.1 Examples of computer crime

Crime	Description
Intellectual property	Infringements of intellectual property are usually civil matters rather than criminal, but the deliberate violation of intellectual property for commercial gain may be a criminal matter
Spam	This is where unsolicited messages are sent to a large number of people usually for commercial purposes such as in advertisements for products, but possibly for fraudulent purposes such as phishing
Phishing	Phishing is the fraudulent activity that aims to obtain sensitive personal information such as user-ids and passwords, bank account and credit card information, etc., by deceiving users that communication is coming from legitimate parties
Identity theft	This is where someone obtains and deliberately uses another person's personal information (e.g., bank account or credit card information) to commit fraud
Credit card theft	This could include the situation where the credit card is skimmed by a skimmer device, and the thieves may then create a cloned card and make purchases. Another approach is the "card not present" fraud where the user's personal credit card information is stolen
Scams	A scam is an illegal or dishonest scheme designed to steal or commit fraud, where the scheme is designed to get the victim part with his or her money. The scam may appear very convincing to the victim
Cyberextortion	This is a crime that involves an attack, or threat of an attack, accompanied by a demand for money to stop the attack
Ransomware	This cyberextortion attack encrypts the victim's data, and threatens to perpetually block access to the data unless a ransom is paid
Cyberbullying	Cyberbullying (or online bullying) is a form of bullying or harassment by electronic means, where someone harasses or bullies another person online (especially on a social media site such as Facebook or Twitter)
Fraud	Fraud is the deception of the victim with the intention of gaining financial gain at their expense. The fraud may be a civil or a criminal matter
Malware	Malware is malicious software that is designed to create harm or cause damage to a computer or network, and they include viruses and Trojan horses
Theft of corporate data	This is where information that is stored on corporate databases or servers is stolen, and it may occur as a result of an employee's account being compromised or due to an unsecured network
Denial of service (DoS)	This attack essentially shuts down a machine or network or where there is a degradation of service that makes it inaccessible to its users, or prevents them from performing productive work
Unauthorised access	Unauthorised access is where someone gains access to a website or server without permission. It may be due to a security breach, where someone gained access by using another person's account, or through guessing his or her password

personal information such as name, address, date of birth, phone number, financial details, login details, passwords for *identity theft*, and the goal may be to use the information gained to access bank accounts, or to sell on the personal information to other criminal groups.

The *cold call* scam is where somebody claiming to be from the technical department of a computer company contacts the recipient, and advises her that their computer is infected with a virus or hacked. They offer to remotely connect to the computer to solve the problem for a fee, and as the victim may be in a state of shock or fear they may well agree to a remote connection. The scammer may simulate a virus on the remote machine and just take the fee, or their actions may be more sinister where they encrypt all of the data on the machine, and demand payment for resolution.

The infamous *Nigeria 419* scam is where the email recipient is offered a share of a large amount of money trapped in the sender's country, if the recipient will help in getting the money out of the country. The recipient may be asked for their bank account details to help them to transfer the money (this information will later be used by them to steal funds), or the request may be to pay fees or taxes to release payment, with further fees requested as time goes on. Of course, the money will never arrive, and *if an email looks like it really is too good to be true then it has a high probability of being a scam.*

12.2.2 Malware

Malware is malicious software that is designed to negatively impact the victim's computer, and it may delete data and software, change user settings, spy on the user, and open up the computer to attacks from those intending to commit fraud. It installs itself on the victim's computer without their consent, and it may install itself without the victim's knowledge by exploiting vulnerabilities in operating systems or browsers, or it may install itself after a user downloads and runs an infected program.

A computer *virus* is a self-replicating computer program that is installed on the user's computer without their consent. This malicious program replicates itself on execution, and infects other computer programs by modifying them. A virus often performs some type of harmful activity on the infected computers such as accessing private information, spamming email contacts or corrupting data. It is not a crime per se to write a computer virus or malicious software. However, if that software or other malware spreads to other computers, then it could be considered a crime.

A *Trojan horse* is a type of malware that is disguised as legitimate software, and it misleads the user on its true intent. Hackers often use this type of software to gain access to the victim's computer system. The origin of the term is from the deceptive Trojan horse that led to the fall of Troy during the Trojan War. The wooden horse contained Odysseus and several other Greeks, and it was left as a victory gift for the Trojans when the Greeks sailed away. The Trojans brought

Fig. 12.1 Trojan horse at Troy

the horse inside their city, and Odysseus and the Greeks later opened the gates of Troy for the returning Greeks, leading to the slaughter of the citizens of Troy and their exodus to found the city of Rome, as described in Virgil's Aeneid. This well-known event in Greek mythology led to the well-known aphorism *"Beware of Greeks bearing gifts"* (Fig. 12.1).

12.2.3 Credit Card Fraud

There are various types of credit card fraud that could occur including "credit card skimming", where a credit card skimmer is a small device that the attacker places on a device where the credit card is swiped. The skimmer collects credit card numbers and information, and the attacker retrieves this information and uses it to create a cloned card and to make fraudulent purchases with the cloned card. It is therefore important to do a quick visual check of a swiped card reader prior to inserting or scanning the card, and while contact less payment technology reduces the risk it does not eliminate it, as skimmers could potentially intercept the data transmitted during a transaction.

Another way in which fraud may occur is where the credit card is lost or stolen, and the person who finds or steals the card uses it to make purchases for the period of time that it remains valid until a block has been placed on the card. Time is of the essence in the case of credit card loss or theft, and so it is essential for the victim to inform the issuing bank as quickly as possible, and to place an immediate block on the card.

Another approach, which is common, for online purchases, is the "card not present" fraud, where attackers steal the user's personal credit card information and use it to make online purchases. In this case the attacker does not possess the physical card, but does possess the credit card number and the card verification value (CVV), and this information is often sufficient to make an online purchase. However, many e-commerce web sites are adopting more stringent security features including third-party verification (TPV), which employs an independent third party to confirm that it is really the customer who is requesting the purchase. For example, the independent third party may send a text to the customer's mobile phone number for verification, or require the customer to use an authentication app on their mobile phone for verification.

Banks have systems for the investigation of credit card theft, and it generally involves examining the transaction data, including time stamps, location data, IP data, and so on to verify first that it was not the cardholder who made the purchases. Most credit card fraud goes undetected and detection rates are extremely low, and often the financial loss is borne by the issuing bank and the merchant rather than the consumer.

12.2.4 Cyberextortion and Ramsonware

Cyberextortion is a crime that involves an attack, or threat of an attack, accompanied by a demand for money to stop the attack. It may involve an individual or group sending a threatening email to a company advising them that they are in a position to seriously harm them, and that they will exploit security vulnerabilities or breaches to launch an immediate attack on the company's computer network unless they receive a payment to prevent the attack. The attack could be to threaten to expose private personal information of customers obtained as a result of a security breach, or it could be a direct attack on the data and information held on databases and servers.

A *ransomware* attack is an even more sinister form of extortion that involves encrypting the victim's files and making them inaccessible, and a ransom payment is demanded to decrypt the victim's data. The victim is unable to recover the files without the decryption key as decryption is an intractable problem, and so the only way to recover the files is to make the payment and pay the ransom. The victim will usually be required to make payment with Bitcoin or another digital currency, and so tracing the perpetrator of the attack is extremely difficult. Once payment is made the victim will be provided with the decryption key, and will be able to access the files.[1]

---

[1] The Health Service Executive (HSE) or Irish Health service suffered a major Ransomware attack in 2021 (during the Covid-19 pandemic) that led to most of the IT systems being shut down. A Russian group called Wizard Spider operating from St. Petersburg is believed to have been responsible

Ransomware attacks are often initiated through malware in an email attachment, which contains a Trojan horse that looks like a legitimate file, and the victim unwittingly opens it. Care is always required when opening an attachment even if it appears to be from a legitimate source, and organisations need good security awareness as well as following good cyber hygiene practices.

The *denial of service attack* is when a web site is overloaded by a malicious attack, and where users are therefore unable to access the web site for an extended period of time. That is, this attack is where the perpetrator attempts to make a computer or network unavailable for use by disrupting the services of a host connected to the Internet. It is achieved by flooding the target computer or server with requests with the goal of overloading the system and preventing normal operation. The attack may be conducted to blackmail the victim for a financial payment, or it may be carried out as an act of revenge.

There are other forms of extortion such as *sextortion*, where the victim is blackmailed into providing sexual favours to the perpetrator, where the latter has obtained indiscreet images of the victim and is threatening to share these on social media. This type of coercion involves the abuse of the power that the perpetrator has over the victim.

12.3 Hacking

A *hacker* is a person who uses his (or her) computer skills to gain unauthorised access to computer files or networks. A hacker may enjoy experimenting with computer technology (the original meaning of the term), but some hackers enjoy breaking into systems and causing damage (the modern meaning of the word). Ethical (*white hat*) hackers are former hackers who play an important role in the security industry in testing network security, and in helping to create secure products and services. Malicious (*black hat*) hackers (also called *crackers*) are generally motivated by personal financial gain, and they exploit security and system vulnerabilities to steal, exploit or sell data (Fig. 12.2).

Many computer systems in use today have vulnerabilities that may be exploited by a determined hacker to gain unauthorised entry to the system, and access to unauthorised information. It is vital that best practice in software and system engineering is employed to develop safe and secure systems, and that known vulnerabilities in system security are addressed promptly by updates to the system software. Further, it is essential to educate staff on security, and to define (and follow) the appropriate procedures to prevent security breaches.

---

for the attack. The attack began with the attackers sending a malicious email, which contained a virus in an attached Excel file, and the file was downloaded allowing the attackers to gain access to the HSE systems. The attackers spent several weeks in the HSE system before they encrypted the data and demanded a ransom demand for the decryption key. It is stated that a ransom for the HSE attack was not paid, and the personal data of several hundred patients was published on line. It took several months to restore the HSE IT systems.

12.3 Hacking

Fig. 12.2 Hacker at work on a blacklit keyboard. Creative Commons

The early hackers were mainly young students without malicious intent who were exploring the university computer systems. These include students at Massachusetts Institute of Technology in the late 1950s that were interested in exploring the IBM 704 computer, and they would enter areas of the system without authorisation and gain access to privileged resources.

They were motivated by knowledge, and wished to have a deeper understanding of the systems that they had access to. The idea of a hacker ethic was formulated in a book by Steven Levy in the mid-1980s [2], and he outlined several ethical principles for hackers, including free access to computers and information and improvement to quality of life.

The *free software movement* was discussed in Chap. 10, and it arose in the early 1980s from followers of the hacker ethic, with Richard Stallman (its founder) often referred to as "the last true hacker" [3]. Today, ethical hackers need to obtain permission prior to acting, as their actions may potentially cause major disruption to an organization. Responsible ethical (white hat) hackers can provide useful information on security vulnerabilities, and they may assist organisations by testing and improving their computer security.

Hackers may probe parts of the system for weaknesses, as system vulnerabilities may allow attackers to gain unauthorized access to the system. Hackers will often attempt to steal confidential data and to disrupt the services being offered by a system. Security engineering is concerned with the development of systems that can prevent such malicious attacks, and recover from them. The security of the system refers to its ability to protect itself from accidental or deliberate external attacks, and it is essential to develop secure systems that can deal with and recover from such external attacks.

There are various security threats in any networked system including threats to the confidentiality and integrity of the system and its data, and threats to the availability of the system. Therefore, controls are required to enhance security and to ensure that attacks are unsuccessful.

The system needs to be designed for security, as it is difficult to add security after the system has been implemented. There is a need to conduct a risk assessment of the security threats facing a system early in the software development process, and this will lead to several security requirements for the system. Security loopholes may be introduced in the development of the system, and so care needs to be taken to prevent these as well as preventing hackers from exploiting security vulnerabilities. We discuss cybersecurity in the next section.

12.4 Cybersecurity

Cybersecurity is the protection of information through good security practices, and it protects the confidentiality, integrity and availability of data. Good cybersecurity is achieved through policies that ensure consistency in employee behaviour in the use of computer technology, as well as training and awareness of security in the workplace. Technology (e.g., firewalls) plays an important role in the implementation of security, and the implementation of good security is achieved through:

- Policy
- Training
- Awareness
- Technology
- Data Security Standards (e.g., ISO 27001).

The system needs to be designed for security, as it is difficult to add security after the system has been implemented. Security threats may be from anywhere, and therefore a holistic approach is required. The protection of privacy requires good security practices to be in place, and these include practices such as:

- Apply need to know principle (only those that need to know should have access)
- Apply minimal user rights principle (level of access should be restricted to the task)
- Update systems regularly
- Design systems with security and privacy in mind.

Security is holistic and it is essential to identify any security vulnerabilities and to correct them. There may be vulnerabilities with the security of the subcontractors of a company, and it is important that their access privileges to the company's computer network be limited. It is important to be able to limit the access that malicious software may have within the company's network, and this may include controls that detect and repel attacks such as shutting down parts of the system or restricting access thereby preventing the malicious software from moving around the network.

Early risk analysis needs to be conducted to determine what needs to be protected, and the threats and vulnerabilities of the current system are analysed as well as their probability and impacts, which leads to a risk profile of the system. The high-risk areas lead to the security requirements including the required security measures and supporting technologies.

Security loopholes may be introduced in the development of the system, and so care needs to be taken to prevent these as well as preventing hackers from exploiting security vulnerabilities. There is a trade off between security risks and the cost of security measures, and this is a continuous process due to continued changes in technology. A comprehensive security system requires a range of measures.

The choice of architecture and how the system is organized is fundamental to the security of a system, and different types of systems will require different technical solutions to provide an acceptable level of security to their users. The following guidelines for designing secure systems are described in [4]:

- Security decisions should be based on the security policy
- A security critical system should fail securely
- A secure system should be designed for recoverability
- A balance is needed between security and usability
- A single point of failure should be avoided
- A log of user actions should be maintained.
- Redundancy and diversity should be employed
- Organization of information in system into compartments.

The cybersecurity policy regulates the behaviour of employees to ensure consistency in what they can or cannot do in order to prevent the misuse of computer resources, or damage or destruction of information. It is essential to develop, implement and manage the cyber security policy, and there may be specific policies for particular systems as well as an organisation wide security policy. The policy defines the vision, sets the direction and scope of security, and provides detailed instructions for its conduct.

The implementation of a good security policy leads to an effective intrusion detection and prevention system, and includes the day-to-day operations risk management, monitoring of security problems, incidence management to handle incidents, disaster recovery management, and business continuity planning. A sample cybersecurity strategy is described in Fig. 12.3.

Cybersecurity training and awareness programs are concerned with educating employees and raising awareness on security. The goal is to ensure the appropriate use of information assets and technology in the workplace. Training helps in improving employee behaviour and in compliance to the security policy.

Technology plays an important role in the implementation of security policy. This may include firewalls that are computer hardware or software that act as a

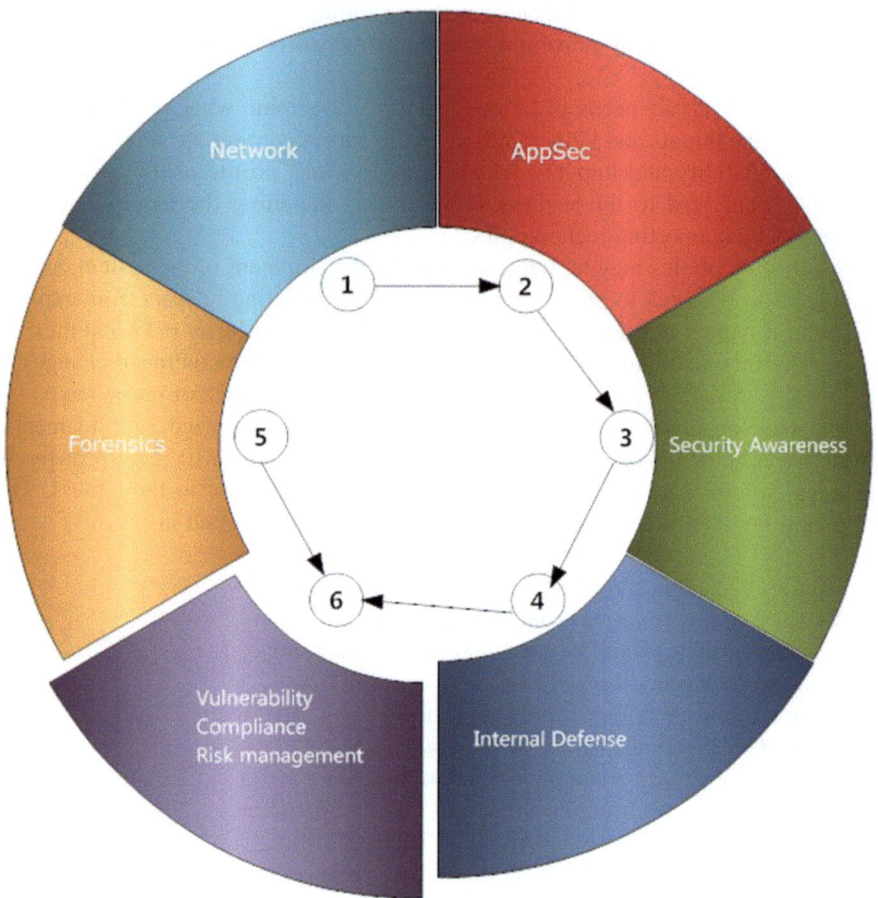

Fig. 12.3 Cybersecurity strategy. Creative Commons

gatekeeper to keep unwanted data out, and prevents unwanted data such as malware from reaching inside the trusted network. That is, firewalls control the flow of information between the outside world (un-trusted network) to the inside world (trusted network). There is a need to employ up to date versions of anti-virus software to protect computers from viruses.

Encryption is a way to reduce system vulnerability, and it involves enciphering the data so that it is unreadable to others who do not possess the security key to decipher it, and so it is an effective way to protect information. A virtual private network (VPN) uses encryption technology to create a secured connection between two points.

Other technical controls include access controls with authentication and authorisation, where the identity of the user is authenticated by password, and a user is assigned a level of access corresponding to her role. There may be controls that detect and repel attacks, and these controls may be used to monitor the system and to take appropriate action to shut down parts of the system or restrict access in the event of an attack. There may be controls that limit exposure (e.g., insurance policies and automated backup strategies) that allow recovery from the problems introduced. Audit logs are maintained to record who, where and when accessed various parts of the system.

Data Security Standards are guidelines for protecting sensitive and confidential information and play an important role in preventing unauthorised access to data. The ISO 27001 standards are a well-known standard for information security, and covers risk management, security controls and security management.

Sun Tzu Wu's description of warfare in "The Art of War" c. 500 B.C. is analogous to security (Fig. 12.4).

- He who knows the enemy and himself will not be in danger in 100 battles.
- He who knows himself but not the enemy will win some and lose some battles
- He who neither knows himself or the enemy will be in danger in all battles.

The security of early computers (large expensive machines) was physical, where access to sensitive locations was protected. The threats in the early days of computing were theft of equipment, espionage and sabotage. The ARPANET network was developed in the late 1960s, and it led to new security challenges, as there

Fig. 12.4 Sun Tzu Wu

were insufficient controls and safeguards to protect data from unauthorised remote users. ARPANET lacked security protocols for phone based connections, and the lack of user based identification and authentication methods meant that hackers had easy access to ARPANET.

ARPANET evolved into the Internet during the 1980s, and TCP/IP became the protocol for ARPANET in the early-1980s. Virtually all computers could now be interconnected by phone line and modem, and the Internet became pervasive in all corners of globe. The Internet brings millions of unsecured computer networks and billions of unsecured computers into continuous communication with each other. Early Internet deployment treated security as a low priority, and many security problems today are due to this lack of security (e.g., email).

The security of information of a particular computer is dependent on the security of every other computer that it is connected to, or another words, there is a clear and present danger from others when the user is on line. This means that there is a need to be conscious of cybersecurity, and to improve security practices to prevent cyber attacks. The goal needs to be to avoid being the victim of information warfare, which could arise if one is undefended. A security breach may be an intentional act by a hostile party, or it could occur accidentally due to a user clicking on an inappropriate link in an email.

The software on a web site runs on the web server, and it may store personal customer data in a database on the hard drive of a server or in the cloud. Hackers will exploit any vulnerabilities to gain access, and the unauthorised activities may include an attack on the data either damaging or compromising information. There have been privacy and security breaches with digital technology allowing personal and confidential data to be compromised. Poor security leads to the theft of confidential data such as credit cards or personal data, identity theft, ransomware attacks, disruption to companies and organisations due to virus attacks, as well as other cybercrimes.

Human error is the major cause of cybersecurity breaches, and so education on security is essential for both employees and the general public. This is to ensure that everyone is aware of the importance of good cyber hygiene practices. Users should choose strong passwords and change them regularly. They should think carefully before clicking on a link or opening an attachment, and to beware of phishing emails that encourage them to give out personal information such as login information for banks.

The World Wide Web consists of unknown users and web sites with unpredictable behaviour operating in unknown countries around the world. These users and web sites may be friendly or hostile, and the issue of trust arises, and one should always err on the side of caution:

- Is the other person whom they claim to be?
- Can the other person be relied upon to deliver the goods on-payment?
- Can the other person be trusted not to inflict malicious damage?
- Is financial information kept secure on the server?

It is essential that the user is confident in the security provided as otherwise they will be reluctant to pass credit card details over the web for purchases. Technologies such as secure-socket layer (SSL) and secure HTTP (S-HTTP) help to ensure security. There is a need for care with technologies such as Wifi as these may not be secure, and there is a need to apply security patches whenever they become available.

Security testing of the software is important, and it is essential to identify security vulnerabilities and any flaws in the security mechanisms of the computer system, and to verify that the security requirements such as confidentiality, availability, and integrity are satisfied. However, the successful completion of security testing does not guarantee that there are no remaining security vulnerabilities in the system, and it is essential to remain vigilant at all times.

There may be legal and privacy violations and negative publicity in the case of security breaches leading to damage to the credibility of the organisation. The unauthorized access to a computer system and the theft of confidential data and disruption of its services is unlawful, and may be subject to prosecution and the full rigour of the legal system of the state.

12.5 Review Questions

1. What is cybercrime?
2. What are the main computer crimes?
3. What is a scam?
4. What is identity theft?
5. What is malware?
6. Explain the difference between cyberextortion and ransomware
7. What is a hacker?
8. What is cybersecurity?
9. Explain the analogy between Sun Wu's description of warfare and cybersecurity.
10. Explain the difference between "white hat" and "black hat" hackers.

12.6 Summary

Computer crime is a crime that involves a computer and a network. The computer may be the vehicle by which the crime was conducted, or it may be the target of the crime. The Internet has dangers with hackers, scammers, and web predators lurking in the shadows. A hacker may be accessing a computer resource without authorisation with the intention of committing an unlawful act including to eavesdropping, or it may be an active man-in-the-middle attack, where the hacker may possibly alter the conversation between two parties.

Many developed countries introduced laws criminalising computer crime from the early 1970s, and new laws have been introduced as technology evolves. The introduction of the Internet and World Wide Web has transformed computer crime to the global stage, with billions of users accessing web sites around the world, and this has created a whole new set of problems and challenges for the legal system to deal with.

Today, more and more individuals and companies are on line, and networking systems and computers have become quite complex. There has been a major growth in attacks on businesses and individuals, and so it is essential to consider computer and network security.

References

1. History of the FBI, in *The Morris Worm.* https://www.fbi.gov/history/famous-cases/morris-worm (1988)
2. S. Levy, *Hackers: Heroes of the Computer Revolution* (O'Reilly Media, 1984)
3. G. O' Regan, *Pillars of Computing* (Springer, 2015)
4. I. Sommerville, *Software Engineering*, 9th edn. (Pearson, London, 2011)

Epilogue

The introduction of modern computer technology has led to immense changes in society. We embarked on a long journey is this book and set ourselves the objective of providing a concise introduction to legal and ethical aspects of computing. The book is written from the point of view of a non-specialist (a software engineer who became interested in legal and ethical issues in computing while employed by Motorola in Ireland), and it is hoped that the reader will find the coverage to be both interesting and insightful.

Chapter 1 introduced ethics and the law in computing, and we discussed ethical and legal problems that arise in the computer field. Chapter 2 presented a short history of ethics, and we discussed ethics in ancient civilisations as well as ethics in several religious traditions. We discussed utilitarian and deontological ethics, and considered the question of what it means to be ethical in a complex world.

Chapter 3 discussed the professional responsibilities of computer professionals, and we discussed the code of ethics of the Association for Computing Machinery, the Institute of Electrical and Electronic Engineers and the British Computer Society. We discussed the role of the whistle blower and its importance.

Chapter 4 discussed ethical software engineering, and we discussed notable failures such as the space shuttle disaster and the defective Therac-25 radiotherapy machine. We discussed the Volkswagen emissions scandal, where engineers designed a "defeat device" that would allow Volkswagen cars to pass their emission tests in the United States.

Chapter 5 discussed the ethics of data science. There has been a phenomenal growth in the use of digital data with vast amounts of data collected, processed and used, and so the ethics of data science has become important. We investigated what is fair and ethical in data science, and what should or should not be done with data. We discussed ethical problems of privacy that arise in social media, the Internet of Things, AI and facial recognition, and the legal aspects of privacy.

Chapter 6 discussed social media and ethics, where social media is designed to have the individual share as much information as possible about themselves, and to continue to do so while they are on the site. Social media sites are designed to be addictive and pose risks to the privacy of an individual. We discussed the Facebook revolution and its impact during the Arab spring, as well as the Cambridge Analytica affair and its impact in influencing voters in the 2016 election in the United States.

Chapter 7 discussed ethics and AI, and we discussed Weizenbaum's Eliza program and the challenge that AI poses to human dignity. We discussed the ethics of self-driving cars, and the need to encode a moral compass to deal with situations where ethical decisions need to be made. We discussed ethical problems that arise with AI and surveillance as well as ethical problems with expert systems.

Chapter 8 provided a brief introduction to law, where the laws of a country apply to all citizens and residents of the state. Roman law influenced the development of civil law, where the civil law tradition is focused on reasoning on the basis of rules. Sharia law acts as a path to guide Muslims in their relationships with their neighbours, the state and with God. English common law operates on the principle of binding precedent, where the judge in a particular case must follow the decision of judges that have made on similar cases.

Chapter 9 discussed legal and ethical aspects of outsourcing of software development and/or testing. The laws and regulations in the country where the subcontractor is based may not be stringent, and so an ethical corporate citizen must do more than just complying fully with the laws of every country in which it is operating, as these laws may not be fit for purpose. We discussed legal aspects of failure as well as lawsuits for breach of contract.

Chapter 10 discussed intellectual property including patents, copyrights and trademarks. Patents protect innovative ideas and concepts, and give inventors exclusive rights to their invention for a specified period or time. A copyright protects the expression of the idea and not the underlying idea itself. A trademark protects names or symbols that are used to identify goods or services.

Chapter 11 discussed e-commerce and e-commerce law, where e-commerce is a way to conduct business online. It often involves listing products for sale in a catalogue format on a website, and there are also legal issues to consider. There is a greater risk of fraud or loss with e-commerce than with traditional commerce.

Chapter 12 discussed computer crime, where the computer may be the vehicle by which the crime was conducted, or it may be the target of the crime. The introduction of the Internet and World Wide Web has transformed computer crime to the global stage, and has created a whole new set of problems and challenges for the legal system to deal with.

We conclude with an epilogue where we summarize the journey that we have travelled in this book.

Index

A
ACM Code of Ethics, 63
Agile development, 72
AI Algorithms and Discrimination, 151
AI and Autonomous Weapon Systems, 153, 156
AI and Disinformation, 152
AI field, 110
Alan Turing, 137
Alexander the Great, 32
Alexandria, 32
Apostasy, 40
Apple Computers, 212
Arab spring, 120
Aristotle, 33
Artificial Intelligence (AI), 137, 139, 161
Association Computing Machinery (ACM), 56
Athenian democracy, 32

B
Balfour declaration, 121
BCS Code of Conduct, 61
Bespoke software, 19
Big data, 99
Bill of Rights Act, 168
Biometric technology, 150
Bridge on the River Kwai, The, 79
British Computer Society (BCS), 56
Buddhism, 41
Buddhist ethics, 41
Business analytics, 98
Business ethics, 3, 4
Business-to-Business (B2B), 224
Business-to-Consumer (B2C), 224

C
Cambridge Analytics affair, 125
Canon law, 172
Chat Generative Pre-trained Transformer (ChatGPT), 158
Chinese ethics, 43
Christian ethics, 35
Civil law, 171
Codes of conduct, 57
Computer crime, 16, 23
Computer ethics, 6
Confucianism, 44
Contracts, 231
Copyright Dispute Apple vs. Microsoft, 212
Copyright for software, 210
Copyrights, 206
Corporate Social Responsibility (CSR), 5, 57
Cyberextortion, 243
Cybersecurity, 246

D
Data analytics, 97, 132
Data collection, 93
Data Privacy Impact Assessment (DPIA), 113
Data science, 93
Deontological ethics, 45, 53
Dieselgate emissions scandal, 82
Digital divide, The, 8
Digital Millennium Copyright Act (DMCA), 211
Digital Rights Management (DRM), 212
Digital Service Act (DSA), 113
Dot com bubble, 225
Draconian law code, 164

E
e-commerce law, 227
Eliza program, 140, 161
English Common Law, 166
Escrow agreement, 189
Ethical decision making, 51
Ethical project management, 78
Ethical project manager, 13
Ethical social media, 133
Ethical software design and development, 79
Ethical software engineer, 12
Ethical software outsourcing, 190
Ethical software tester, 13
Ethical software testing, 85
Ethics, 2, 29
Ethics of data science, 92
European Convention on Human Rights (ECHR), 168, 176
European law, 175
Expert system, 138, 145, 161

F
Facebook, 118, 119, 130
Facebook revolution, 119
Facial recognition, 110, 149, 161
Fake news, 131, 135
Free Software Foundation (FSF), 215

G
GDPR law, 90
GDPR privacy law, 112

H
Hacker, 25, 244
Hacking, 25, 244
Hammurabi code, 30
Hellenistic age, 32

I
IEEE Code of Ethics, 61
Immanuel Kant, 46
Informed consent, 92
Instagram, 118
Institute of Electrical and Electronic Engineers (IEEE), 56
Intellectual property, 16
Intellectual property law, 197
Internet of Things (IoT), 108
Islamic ethics, 38

J
Jack Dorsey, 128

L
Law of tort, 21
Lawsuits in computing, 22
Legal aspects of computing, 14
Legal aspects of outsourcing, 19
Legal aspects of trademarks, 217
Legal breach of contact, 192
Legal impacts of failure, 18
LinkedIn, 118, 130

M
Machine learning hiring algorithm, 152
Magna Carta, 167
Malware, 241
Model, 72
Moderating content, 127

N
Natural law, 178
Natural rights, 177
Nicomachean ethics, 34
Nigeria 419 scam, 241

O
OKCupid, 93
Outsourcing, 183, 184

P
Panopticon, 101
Parnas, 11, 71
Patent litigation, 203
Patent ruling on first computer, 204
Patents, 198
Phishing, 239
Plato, 49
PLATO system, 118
Precautionary principle, 66, 68, 77
Predictive policing, 152
Privacy policy on website, 230
Professional engineers, 12, 56, 71
Professional ethics, 3
Professional responsibility, 12, 55, 56
Project, 78
Project management, 78
Protagoras, 3

Index

R
Ransomware, 244
Rational Unified Process (RUP), 72
Registering a domain name, 229
Remote project management, 184
Request for Proposal (RFP), 187
Right to have privacy, 101
Risks and issues with AI, 158, 159
Robots, 155
Rossums Universal Robots, 138

S
Safety and ethics, 73
Safety critical system, 73
Scam, 239
Searle's Chinese Room, 137
Security, 25, 114, 245
Security loopholes, 247
Self-driving car, 141, 144
Separatism, 77
Sharia law, 173
Social media, 107, 117, 253
Social media campaign, 124
Socrates, 32
Software engineering, 69
Software license, 214
Software licensing, 17
Space shuttle challenger disaster, 76, 87
Spiral model, 72
Statement of work, 189
Steve Jobs, 219
Stoicism, 35
Super-intelligent machines, 156
Supplier selection, 184
Surveillance, 149
Syllogistic logic, 35

T
Therac-25 disaster, 74
Trademark Dispute Apple Corps vs. Apple Computer, 219
Trademarks, 216
Treaty of the European Union, 175
Treaty on the Functioning of the European Union, 175
Tripartite system, 169
Trojan horse, 241
Trolley problem, 144
Turing test, 139
Tweet, 128

U
UN Declaration of Human Rights, 167
Universal Declaration of Human Rights (UDHR), 178
Utilitarianism, 44

V
Value Centred Design (VCD), 81
Virtue ethics, 48
Volkswagen, 84
Volkswagen emissions scandal, 83
Vulgate, 206

W
Waterfall model, 72
Weizenbaum, 141
Whistleblower, 58, 59

MIX
Papier aus verantwortungsvollen Quellen
Paper from responsible sources
FSC® C105338

If you have any concerns about our products,
you can contact us on
ProductSafety@springernature.com

In case Publisher is established outside the EU,
the EU authorized representative is:
**Springer Nature Customer Service Center GmbH
Europaplatz 3, 69115 Heidelberg, Germany**

Printed by Libri Plureos GmbH
in Hamburg, Germany